WITHDRAWN
USJ Library

The River Book

©1998 Connecticut Department of Environmental Protection

Painting on cover: "Autumn Morning"
© 1998 Robert T. Blazek, A.S.M.A., Litchfield, Connecticut
Oil on canvas; use for cover donated by Robert T. Blazek

MacBroom, James Grant.
The river book: the nature and management of streams in glaciated terranes / by James Grant MacBroom.
p. cm. — (DEP bulletin; 28)
"Connecticut Department of Environmental Protection, Natural Resources Center."
Includes bibliographical references and index.
IBSN 0-942085-06-X (alk. paper)
1. Rivers. 2. River engineering—Environmental aspects.
I. Connecticut. Dept. of Environmental Protection. II. Title.
III. Series.
GB1203.2.M23 1998
551.48'3—dc21 98-29653
CIP

Published by the DEP Natural Resources Center
Technical Publications Program
79 Elm Street, Store Level
Hartford, Connecticut 06106-5127
For ordering information phone (860) 424-3555

ISBN 0-942085-06-X

The Connecticut Department of Environmental Protection is an equal opportunity agency that provides services, facilities, and employment opportunities without regard to race, color, religion, age, sex, physical or mental disability, national origin, ancestry, marital status, or political beliefs.

The River Book

The Nature and Management of Streams in Glaciated Terranes

by James Grant MacBroom

NATURAL RESOURCES CENTER

CONNECTICUT DEPARTMENT OF ENVIRONMENTAL PROTECTION

DEP BULLETIN 28

This book was funded in part by grants from the Federal Emergency Management Agency and the Long Island Sound License Plate Program

CONNECTICUT DEPARTMENT OF ENVIRONMENTAL PROTECTION

The Honorable John G. Rowland
Governor of Connecticut

Arthur J. Rocque, Jr.
Commissioner of the Department of Environmental Protection

David K. Leff
Assistant Commissioner
of the Department of
Environmental Protection

Steven O. Fish
Director of the
Natural Resources Center

Allan N. Williams
Publisher

David P. Wakefield, Jr.
Text and Copy Editor

Michael Oliver, Ph.D.
Julie Tamarkin
General Editing

Acorn Studio, Hartford
Book Design

Roberta J. Buland
Indexing

Table of Contents

Figures, Photographs and Tables

Figures

Figures, continued

Photographs

Photographs, continued

Tables

Tables, continued

Foreword

Is it possible to find a Connecticut scene like that of this book's cover? The answer is a reassuring, yes. There are many such places. Still, the existence of a stream today does not guarantee its future, or its future as a reservoir of appreciated uses. Yesterday, naturally flowing water was treated as if it were plumbing, confined flows within predictable environments, easily altered for the convenience of some. Today, natural systems are more likely to be analogous to the inter-connectiveness within the human body than to the plastic piping in our homes.

River protection comes from a society that values clean water, honors diverse species, and treasures the benefit to the soul of flowing places. A river's roots extend not only to its channel, banks, and floodplains, but virtually to all land. A river does not begin, it continues. It continues the elaborate cycle of water falling upon the earth, running over and through it, evaporating from it, and moving downslope to other water bodies, there to be evaporated and start the cycle over. I hope this book conveys the awe-inspiring complexity of our streams, and from that understanding the landscape and its streams will benefit.

This book is not for everyone, but almost. The fluvial geologist, the stream ecologist, the hydrologist, and the civil engineer might search in vain for formulas and greater detail. The water quality analysts, fisheries professionals, and wetland scientists might wish to add more to their subjects. Yet, the stream ecologist might find of interest information on channel dynamics, and the engineer might find useful the chapters on impacts. The town planner, wetland commissioner, property owner, and land developer will find much to ponder. The science teacher might delight at finding a book whose text is relevant to the student's landscape. For those whose actions will affect rivers, this book provides a better understanding of the natural and human

factors that affect our rivers.

There are those who ask if this book is strictly about Connecticut's rivers. It certainly has been authored and published in Connecticut, and the majority of examples and illustrations are from Connecticut. Yet, similar climate and natural resources may be found throughout any area where glaciers have been, where there are general similarities in landscape form, and where humans have generally repeated similar modifications of rivers. All the numbers may not be the same, but the processes are. To that extent this book is about much of the land in the world's northern climates.

The author's original charge was to write a pamphlet on flood management. To his credit, he did not stop there. Paragraphs became chapters, and chapters multiplied as the author realized the flooding of a river is inseparable from all other components. The author and his professional colleagues donated hundreds of hours, many evenings and weekends, for which the reader should be grateful.

I invite the reader to use and enjoy this book. Most of all, I enjoin all to protect our flowing birthright.

Allan Noam Williams
Publisher

Preface

Other than an ocean, there is probably no natural feature more prominent in the landscape of this water planet, or upon the human consciousness, than that of the flowing river.

As dependent as humankind is on fresh water and all that it nurtures, it is not surprising that patterns of human settlement, agriculture, and industry have revolved around, and have evolved with, the streams and rivers that comprise the aquatic arteries of the terrestrial world.

Yet, since the dawn of civilization, the sustenance our species has derived from these waters has been tempered by the overwhelmingly destructive force of floods, against which, until quite recently in history, humans had little defense save prayer.

It is only since the beginning of the Industrial Era, and especially the past 150 years, that mankind's technological ability has achieved the power and scale sufficient to rival nature as a force in shaping the land. During much of that era, most humans viewed nature as a force to be tamed, considered the Earth to have inexhaustible resources, and saw the environmental impact of one individual as virtually inconsequential.

Today, however, as scientific knowledge of the interrelation of natural systems catches up with mechanical wizardry, it has become apparent that human manipulation of natural waterways has consequences not always understood or foreseen. As our power to alter the landscape increases and the world's population grows, the cumulative effect of what once seemed insignificant activities has reached a scope that has the power to degrade the natural systems around which civilization is based.

At the same time, even as larger and larger segments of society are concentrated in urban centers, increasing numbers of people are realizing what has been lost in the divorce of our cities, and our persons, from their natural settings.

Never has it been more important to understand how natural systems work and how human activities affect them. And in a democra-

cy where much land use is regulated by boards composed of citizen volunteers, it is imperative that such knowledge be accessible not only to scientific specialists, but to laymen as well.

Today, literally every corner of the Earth is affected in some measure by human activity. From the decision to preserve a wilderness stream, to the mitigation of flooding in populated areas, to the management of commercial navigation on a major river, the future of virtually every waterway is in some way dependent on our global culture. As we assume this important responsibility, it is our obligation as citizens, regulators, and lawmakers to be as well-informed as is possible.

David P. Wakefield, Jr.
Editor

Acknowledgements

I would like to express my appreciation to the many people who participated in this project. Allan Williams deserves special credit for his guidance and patience, as do members of the Department of Environmental Protection River Advisory Committee, with whom we discussed many parts of the book.

The following people reviewed and commented on one of the many drafts of the manuscript. They provided invaluable comments on the organization and content of this publication.

Of present and former employees of the Department of Environmental Protection, I wish to thank Pamela Adams, Peter Babcock, Fred Banach, Charles Berger, Jr., Doug Cooper, Lee Dunbar, Charles Fredette, Steve Gephard, Bob Hartman, Carolyn Hughes, Bill Hyatt, Rick Jacobson, Chuck Lee, Al Letendre, Alan Levere, Ralph Lewis, Art Mauger, Kenneth Metzler, Jim Murphy, Jay Northrup, Ernest Pizzuto, Wanda Rickerby, Ron Rozsa, and Steve Tessitore.

From a host of other organizations, I wish to thank Christine Adams, Regional Water Authority; Cindy F. Alleman, Connecticut Water Company; Nick Bellantoni, State Archeologist; Claire Benoit, Branford, Connecticut; Reeve Biggers, Redding Conservation Commission; Judith Cantwell, Connecticut Department of Transportation; Peter Curry, the Metropolitan District; Tim Dodge, USDA Soil Conservation Service; John Friar, Middlefield; William Haase, Ledyard town planner; Joe Neafsey, USDA Soil Conservation Service; Charles Paino, Lake Lillinonah Authority; Alan Ragins, National Park Service; Pam Ritter, Mill River Wetland Committee; Joy Shaw, Mill River Wetland Committee; Carol Szymanski, Capitol Region Council of Governments; Dr. Karl Eric Tolonen, New Haven.

Additional contributions came from Penelope Sharp, Edward Hart, Laurie Giannotti, Jeanine Bonin, and Vincent C. McDermott. The drawings were prepared by William Nagle, Rodney Shaw, Colleen Nanna, Laura Wildman, and Sarah Moore.

Ralph Lewis brought to Chapter Two (Channels and Floodplains) the latest understanding of our glacial history. Ken Metzler provided a constructive review of Chapter Three (Ecology), and Bill Hyatt contributed to the aquatic habitat material. Ron Rozsa provided extensive assistance in Chapter Five (Tidal Rivers and Marshes). The original draft was edited by Linda Case; David Wakefield did the final copy rewrite; and Michael Oliver and Julie Tamarkin proofread the text. Mary Crombie did the graphic design and prepared the manuscript for publication.

Jay Northrup and Chuck Berger, both responsible for flood management and the original sponsors of the project, listened in good faith and agreed to the numerous requests to change the scope and schedule of the book. Over 100 people have contributed to or reviewed this book, and many, especially early on, had doubts that such a range of topics could be integrated and presented in one volume. They all participated because of the need for such a compilation. I thank all the reviewers for their trust in us to produce this book. I respect their contributions of time and their dedication to the protection of rivers. The same dedication and desire is displayed in the magnificent cover painting, created and donated for this book by Robert Blahzek of Litchfield.

James Grant MacBroom

This book is dedicated to

my mother, who taught me to appreciate nature;

my father, who stimulated my interest in engineering;

my family, Valerie, Skye, and Tyler, who share my life;

my colleagues at Milone and MacBroom, Inc., who contributed to this book;

my teachers, instructors, and professors, who shared their time and knowledge; and

those who strive for a sustainable environment.

We must integrate the sciences in our studies of rivers.

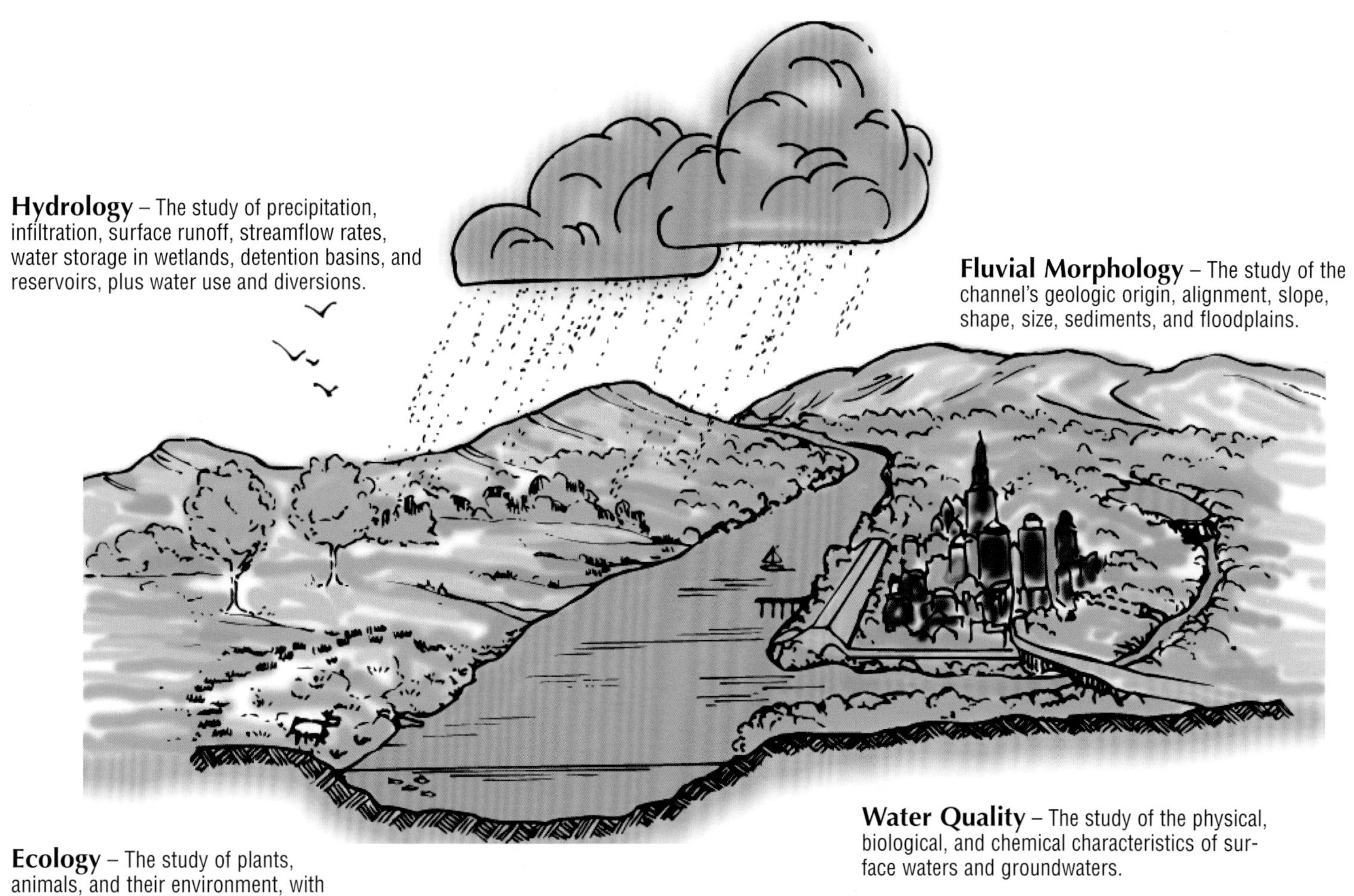

Hydrology – The study of precipitation, infiltration, surface runoff, streamflow rates, water storage in wetlands, detention basins, and reservoirs, plus water use and diversions.

Fluvial Morphology – The study of the channel's geologic origin, alignment, slope, shape, size, sediments, and floodplains.

Water Quality – The study of the physical, biological, and chemical characteristics of surface waters and groundwaters.

Ecology – The study of plants, animals, and their environment, with emphasis on aquatic systems, wetlands, and riparian forests.

Hydraulics – The study of the stream's water velocity, flow depth, flood elevations, channel erosion, storm drains, culverts, bridges, and dams.

TYPICAL WATER-RELATED FEATURES FROM HEADWATERS TO THE OCEAN

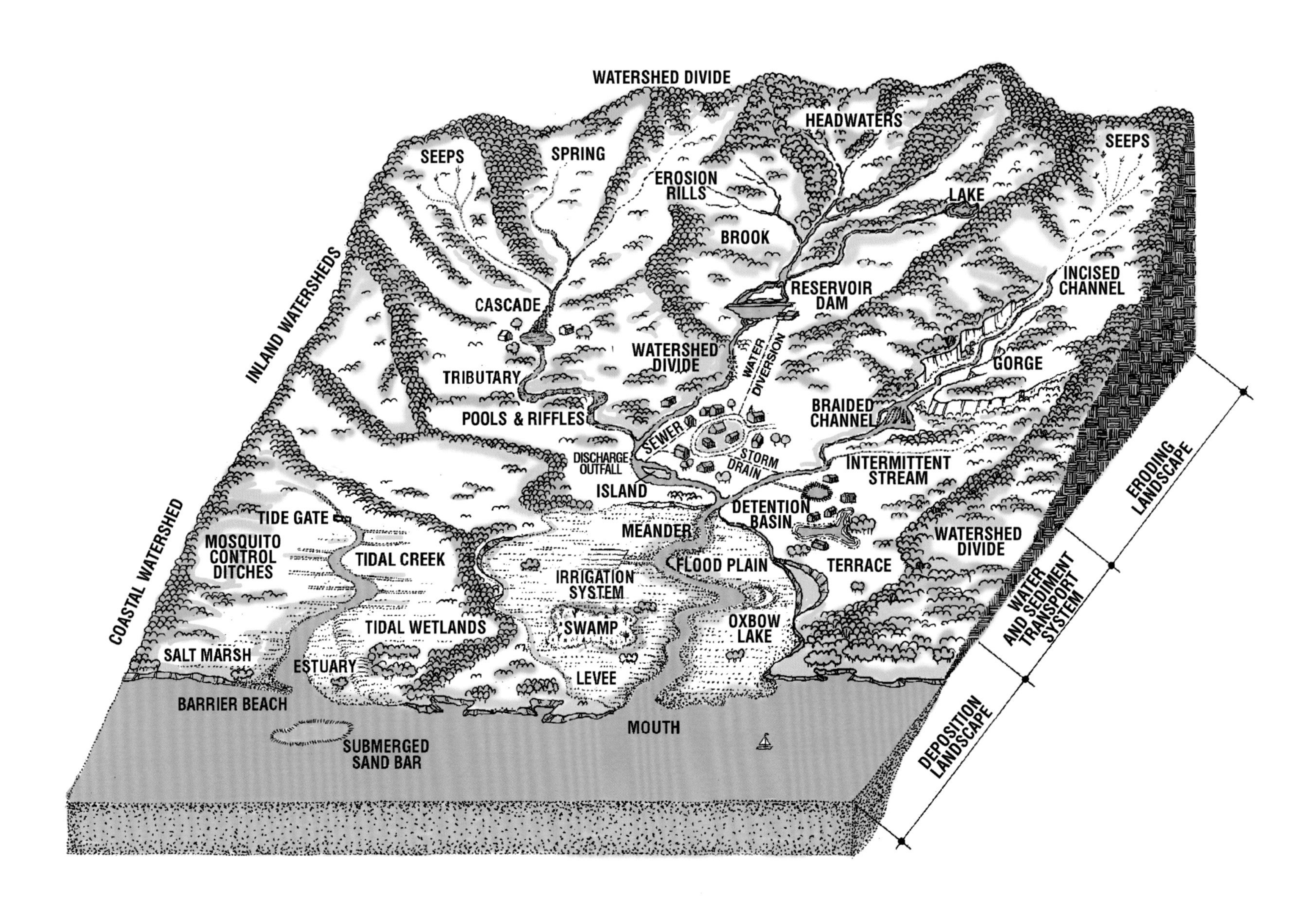

Modified from the concept presented by Laura Lee Lizak as found in *The World Book Encyclopedia*.

The River Book

The Nature and Management of Streams in Glaciated Terranes

STREAM PROCESS INTERRELATIONSHIPS

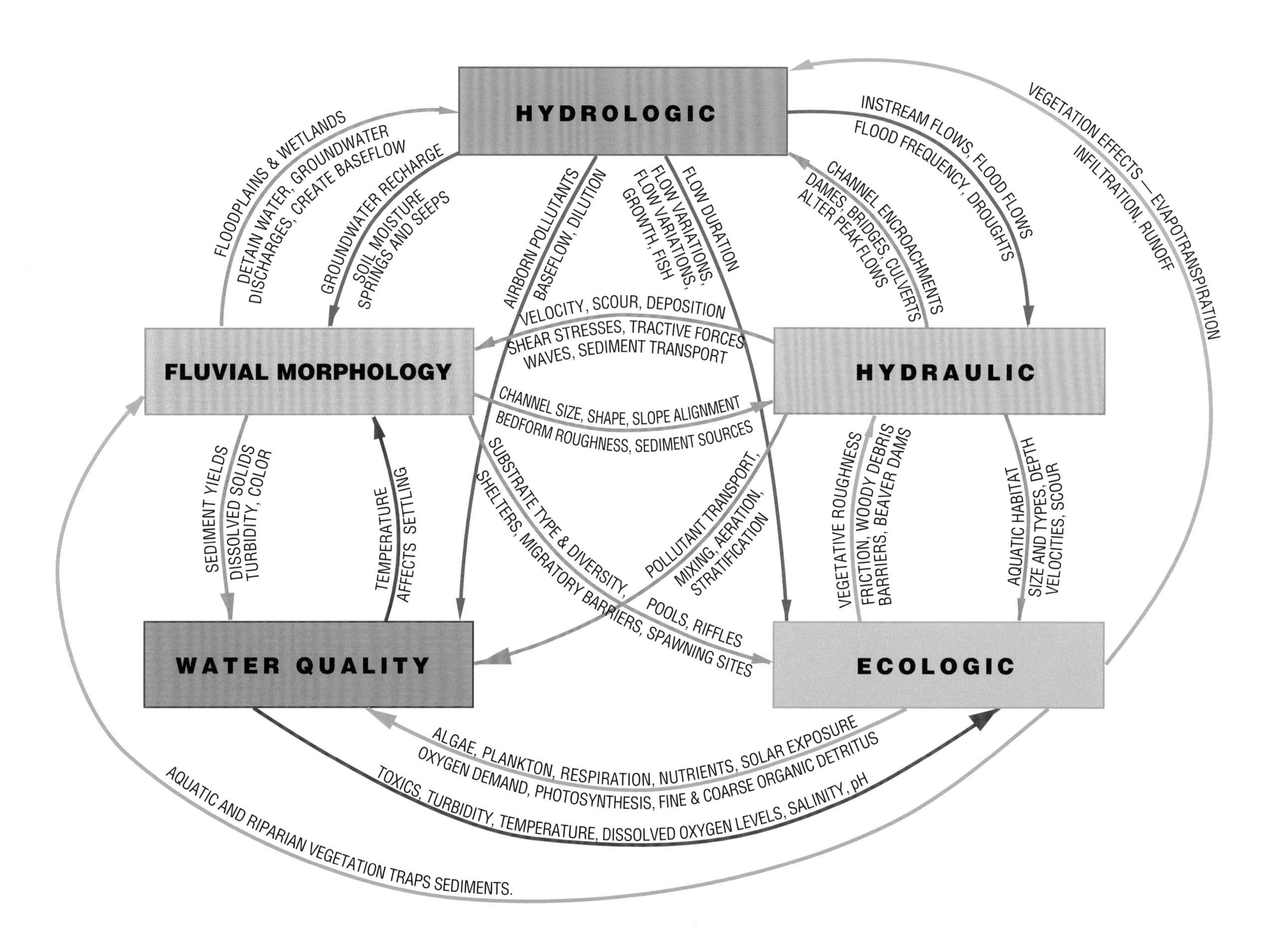

How Rivers Are Used

Rivers vary from natural, rapid-flowing mountain streams to busy city waterfronts lined with factories, highways and railroads. Many people ignore rivers except during extreme natural events. Others think of rivers only as habitats for fish and other wildlife, and as sites for peaceful outings. However, as in the past, waterways are essential to the health and economic prosperity of humans and their communities.

Historically in New England, rivers were essential not only as sources of food and water, but also for commerce. Shellfish, finfish and waterfowl were principal sources of food for early settlers, as they were for Native Americans. And transportation and trade were the lifeblood of early communities.

Streams were vital in the timber industry, which transported logs to mills in yearly drives. The first gristmill was built in 1637, and there was a time when virtually every swift-flowing river generated water power for sawmills, gristmills and textile factories. By the mid-1800s, even before the era of large water-powered enterprises, Connecticut had more than 200 small mills.

After the Civil War, extensive water-powered textile mills in eastern Connecticut were using canals, reservoirs and water wheels. Many large mills became the centers of entire communities.

In western Connecticut, rivers were harnessed for the huge brass industry, sparking prosperity throughout the Naugatuck River Valley. The power was used for forging, washing, and cooling at mills in Torrington, Watertown, Waterbury, Seymour and Ansonia.

One of the first rivers with an integrated electric power generation system was the Housatonic. The Falls Village Station at Great Falls was built in 1912 as a run-of-the-river plant, one that simply harnesses the power of the flowing water, followed by construction of the Rocky Ridge Station, the first large pump storage plant in the coun-

What Rivers Do

- Supply water
- Assimilate waste
- Provide aquatic habitat
- Convey runoff
- Transport goods and people
- Serve as a source of food
- Provide recreational activities
- Generate hydroelectric power
- Irrigate lands and crops
- Help in industrial processes

An example of a complex mix of river and city.

try. The station uses excess electricity to pump river water up to Candlewood Lake. This water is released during the hours of peak power demand to turn turbines and generate electricity. The Housatonic system, still in operation today, also includes the plants at Shepaug and Stevenson Dams. They form Lake Lillinonah and Lake Zoar, which are also used for recreation.

Today, for the most part, cities and towns rely on rivers primarily for water supply, drainage, and conveyance of treated waste water. Impounded rivers form reservoirs that are the main source of drinking water for all of the larger cities and many suburbs. In some areas of the state, river water irrigates field crops.

Some electricity is still produced by hydroelectric generators. We depend on river water to cool power generation plants, and industries use it to carry waste and for cooling. Rivers are still used to transport goods such as oil, coal, gravel, and bulk commodities.

These activities are essential to the economy. Nevertheless, they may conflict with other needs — human, recreational, ecological. The many recreational uses for rivers include swimming, fishing, boating, rafting, tubing, canoeing, kayaking, camping, and hunting. And obviously, rivers add to scenic beauty and help make Connecticut an attractive place to live.

Often it is impossible for a single stream to fulfill conflicting needs. Which is more important — to be able to have safe water to drink, or to have a means of disposing of treated waste products?

In this era of fast transit and rapid communication, just as in Colonial times, it is desirable for a community to be located on a river. But in attempting to meet our every need, we frequently confront the irony of degrading the river upon which a community's identity was founded.

Given our growing population and the limit to our natural resources, it is not surprising that conflicts have arisen. We have no other choice now but to manage and preserve our rivers to meet competing needs.

The task, then, is for decision makers to weigh all factors and develop sound solutions for managing our river resources.

One of many factories that formerly used river water power for manufacturing.

Major Connecticut Textile Mills, Past and Present, and Their Power Sources

Mill	Power Source
Pomfret	Quinebaug River
Control Village	Moosup River
Sterling	Quinebaug River
Danielsonville	Quinebaug River
Putnam	Quinebaug River
Baltic	Shetucket River
Willimantic	Willimantic River
Wauregan	Quinebaug River
Taftville	Shetucket River
Manchester	Hop Brook
Rockville	Hockanum River
Jewett City	Pachaug River

Surface Water

A river seems a magic thing. A magic, moving, living part of the very earth itself.

Laura Gilpin, *The Rio Grande*, 1949

In urban areas, riverside parks offer residents a variety of recreational opportunities, such as this walk at Charter Oak Landing in Hartford, on the Connecticut River.

Surface Water

Rivers and other streams are essential to our lives. They provide water to drink and to irrigate cropland; they power our industries, provide transportation, and help treat and assimilate our wastes. They support wildlife and serve as sites for recreation.

Yet despite our dependence on waterways, we have incomplete knowledge of how their physical, chemical, and biological processes interact. There is a delicate balance between a river's hydrology and hydraulics, morphology, ecology, and water quality that must be maintained.

It is important to recognize that although Connecticut's water resources are abundant, they are not unlimited. It is our responsibility to conserve and manage them for long-term use. Unless we do so, we risk not only degrading them, but also diminishing the quality of our lives.

Increased knowledge can but help in the management of these critical resources, assuring their benefits for generations to come.

TABLE 1

OFFSTREAM AND INSTREAM USES OF RIVERS

OFFSTREAM USES	INSTREAM USES	
CONSUMPTIVE USES	NON-CONSUMPTIVE USES	RECREATIONAL AND CULTURAL USES
Crop irrigation	Hydroelectric power	Whitewater sports
Water for livestock	Other power-generation plants	Boating
Off-site cooling	On-site cooling	Fishing
Off-site wash water	On-site wash water	Tubing
Water-supply diversions	Hydromechanical power (mill power)	Swimming
Aquifer recharge	Transportation (navigation)	Aesthetics

WHAT IS A RIVER?

A stream is a body of water that flows downhill following a channel, a natural or artificial depression in the land that confines the water within a definite bed and banks. We term the larger ones rivers; they drain areas of significant size and usually maintain a continuous, perennial flow.

Brooks and creeks are streams generally considered to be smaller than rivers, though there is no precise delineation. The term "brook" is applied to small, shallow streams of fresh water, ususally with a somewhat turbulent flow, while "creek," in the East, usually refers to streams of moderate size draining a small area, often describing tidal channels in coastal regions. The smallest streams, especially those having poorly defined channels, are called rivulets or streamlets.

The term "watercourse" is sometimes used interchangeably with

Rivers are put to many uses in urban areas.

"stream," while some people also apply it to lakes, ponds, marshes and other bodies of water connected to or by streams.

Perennial streams are those that flow throughout the year. Those that do not are classified as intermittent. These generally are found in the higher regions of watersheds, where there is limited runoff and groundwater discharges, and can sometimes disappear entirely into permeable soils. In the Northeast, they usually drain areas of less than 100 acres.

WATERSHEDS

Understanding watersheds is crucial to the study of rivers. Water movement through a watershed begins with runoff that flows downhill over the ground as sheetflow, then collects in small rills or rivulets that erode shallow channels in the soil, and feed small streams. These streams receive more runoff and groundwater discharge as they descend, eventurally merging where their valleys intersect. In large watersheds, they join to form major rivers that ultimately empty into the oceans.

Rivers provide vital locations for recreation and play.

The characteristics of the surface flow in a particular watershed are determined by hydrological factors that include precipitation, temperature, soil moisture and the length of the growing season; and by physical factors such as topography, soil and bedrock types, fracture patterns, the rate at which water infiltrates into the ground, land use and vegetation.

The pattern of southward drainage that developed in most New England watersheds was determined by the control that the bedrock and glacial deposits exerted on streams as the last glacier melted away, and the post-glacial topography emerged. This topography was shaped over the past 500 million years as continental collisions and rifting folded and fractured the bedrock to create a general north-south alignment of rock types and fracture systems.

Pre-glacial streams had millions of years to preferentially remove less resistant rock types, and the broken rocks in fracture and fault zones. The pattern of this differential erosion was eventually defined by the resistant north-south trending rocks of the hills, and intersections of the less resistant rocks and fractures of the valleys. The two known glaciers that moved from north to south across New Engalnd enhanced the existing bedrock topography by rounding the hills and overdeepening the valleys.

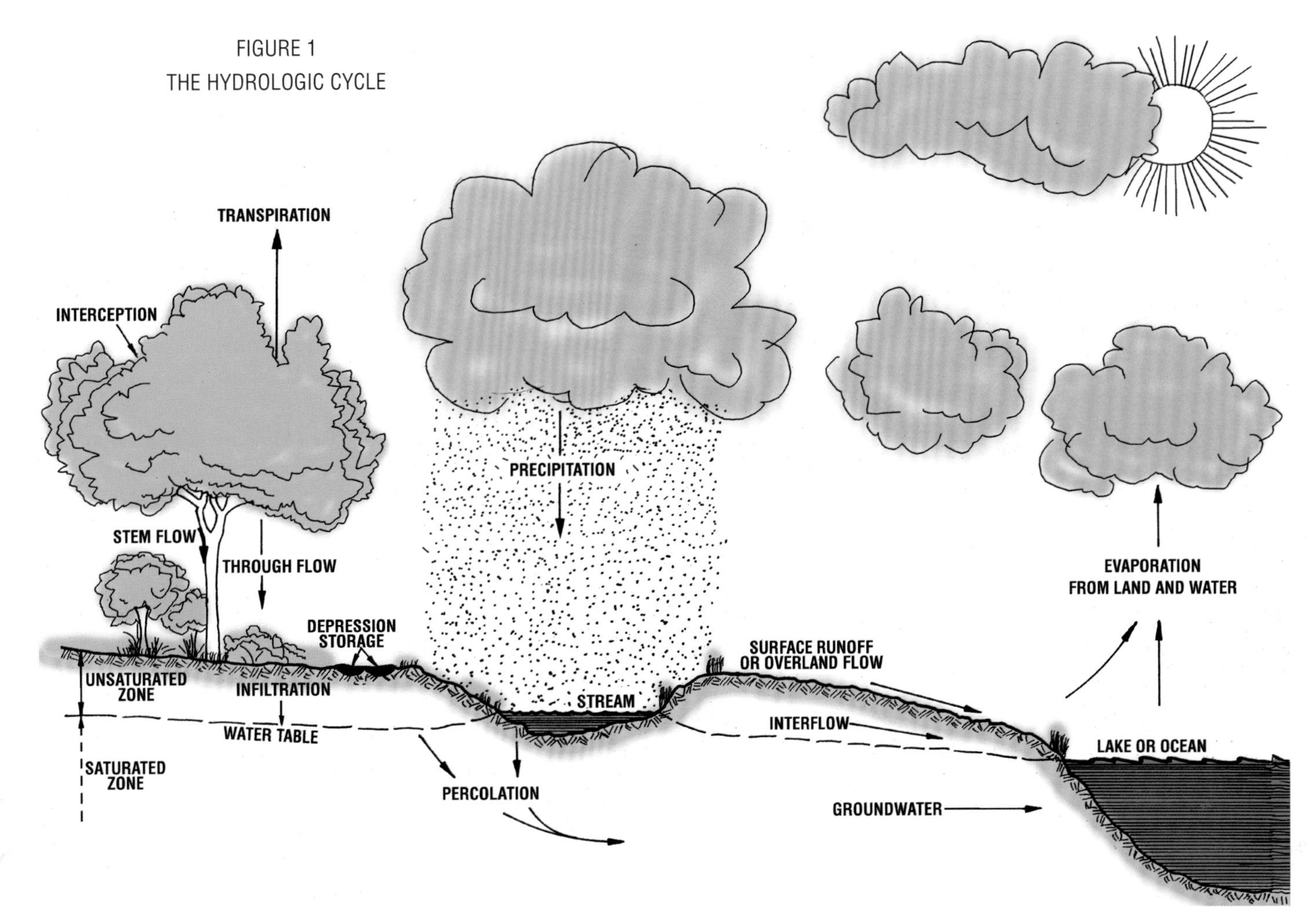

As the last glacier advanced and melted away it left unsorted material known as till exposed on the hills, and it filled the valleys with better sorted clays, sands, and gravels known as stratified drift. In most cases, as the ice melted, the emerging till-covered hills confined the streamflow and directed it to existing bedrock valleys.

The results of this "inherited" topographic control on stream courses and watershed configurations are still evident. Most modern streams in New England originate in the till-covered hills and eventually flow across stratified drift in the valleys. The orientation and postion of most of these streams, and the overall drainage pattern of their watersheds, is related to rock type and/or fracture occurrence, and most of central New England drains south to Long Island Sound.

TABLE 2

MEAN MONTHLY AND ANNUAL WATER BUDGET FOR THE FARMINGTON RIVER BASIN

MONTH	PRECIPI-TATION	EVAPOTRANS-PIRATION	PRECIPITATION MINUS EVAPOTRANS-PIRATION	RUNOFF	CHANGE IN STORAGE
October	3.74	1.72	2.02	1.22	0.80
November	4.56	.73	3.83	1.89	1.94
December	3.97	.20	3.77	2.01	1.76
January	3.91	.20	3.71	2.38	1.33
February	3.25	.20	3.05	2.05	1.00
March	4.31	.38	3.93	4.10	-.17
April	4.04	1.50	2.54	4.32	-1.78
May	3.98	2.84	1.14	2.62	-1.48
June	4.17	3.80	.37	1.67	-1.30
July	3.99	4.46	- .47	1.02	-1.49
August	4.35	3.96	.39	1.15	-.76
September	4.15	2.98	1.17	1.20	.15
Mean Annual	48.42	22.97	25.45	25.45	0

(ALL VALUES IN INCHES OF WATER)

Source: United States Geological Survey

HOW WATER CIRCULATES — THE HYDROLOGIC CYCLE

The natural system by which water circulates through the Earth's atmosphere, over its surface and beneath the ground is called the hydrologic cycle.

Water vapor enters the atmosphere when the sun's heat causes it to evaporate from oceans, lakes and streams, as well as directly from snow, ice and soil; and through transpiration, the release of water vapor by plants during photosynthesis.

Warm, moisture-laden air rises in the atmosphere until it cools to the point where the water vapor condenses and falls as rain, snow, sleet or hail.

Some precipitation evaporates before it reaches the ground. Of that which does reach the surface, frozen precipitation accumulates until the temperature rises sufficiently to melt it, releasing it back into the hydrologic system. Rainfall continues its journey immediately. Precipitation and evaporation vary on a seasonal basis (see Table 2, Page 8).

Of the precipitation that falls as rain, a portion lands on lakes and streams.

In forested areas, much of the rain is intercepted by trees and shrubs, more so during the growing season than during the leafless stages of fall and winter. The rest passes through the vegetation canopy and lands on the ground; this is called throughfall.

The canopy reduces and delays runoff. The rain falling on it may evaporate, especially after light storms. Water may drip from leaves, striking the ground with less force than direct rainfall. Water may reach stems, branches and trunks, where some will be absorbed by the bark and may be used by attached organisms such as lichens. Significant amounts may flow down the trunk to the ground, a phenomenon known as stemflow.

Depending on soil conditions, water that reaches the ground may infiltrate into the soil or remain on the surface. Infiltration rates depend on soil texture and structure, pre-existing moisture content, and cover by vegetation and impervious material. Loose, sandy soils can absorb more water than dense, clay-type soils; dry soils more than wet. Infiltration rates also are affected by soil compaction and, during cold weather, frost and snow cover.

Some of the infiltrated water returns to the atmosphere through evaporation; some is taken up by the roots of plants and is eventually transpired into the air. Some water adheres to soil particles and is held as soil moisture. Other water is drawn by gravity down to the deeper groundwater, while some flows downhill in the top layers of the soil as interflow (often also referred to as subsurface runoff), discharging through seeps and springs into streams and wetlands. (See Figure 1.)

The excess rainfall that does not infiltrate the soil accumulates on the surface, filling depressions and flowing downhill in thin layers called sheetflow. Sheetflow is most noticeable during intense rain. In densely developed areas, the removal of vegetation and the presence of impervious surfaces such as buildings and pavement increase sheetflow and accelerate its movement to rivulets, streams and rivers.

The hydrologic cycle can be thought of as a "water budget" made up of receipts, disbursements and assets on hand. Connecticut's "receipts" would include precipitation falling in the state and water flowing in from Massachusetts, Rhode Island and New York. "Disbursements" are losses through evaporation and plant transpiration (known together as evapotranspiration) and outflows to other states and the sea. "Assets on hand" comprise the water on the surface and in the ground in Connecticut. Though various entries change from year to year, the water budget always balances. (See Table 2.)

In Connecticut, precipitation usually is fairly uniform throughout the year. However, evapotranspiration, which ranges from the equivalent of 21 to 24 inches annually, is proportional to temperature and fluctuates widely with the seasons. Thus, by late spring, as temperature increases and plants end their winter dormancy, there generally is a moisture deficiency in the soil and less water is discharged into streams. As temperatures decline in the fall and plants become inactive, soil moisture levels usually increase and groundwater is recharged.

Marinas, docks, walkways, and parks invite people to use the river recreationally.

Precipitation

During the past century, there has been a statistically significant trend of increasing annual precipitation. At the present, precipitation in Connecticut varies from an average of 48 inches a year along the southwest coast to an average of 52 inches in the northwest hills. Most falls as rain; the rest as snow, sleet, hail and freezing rain.

Hydrologists measure precipitation in several ways, including depth, duration and intensity.

Depth is a measure of the amount of rainfall (or the melted equivalent of frozen precipitation) that falls during a specific storm or over a given period of time. It is expressed in terms of the vertical depth in inches or millimeters.

FIGURE 2

EAST COAST HURRICANES

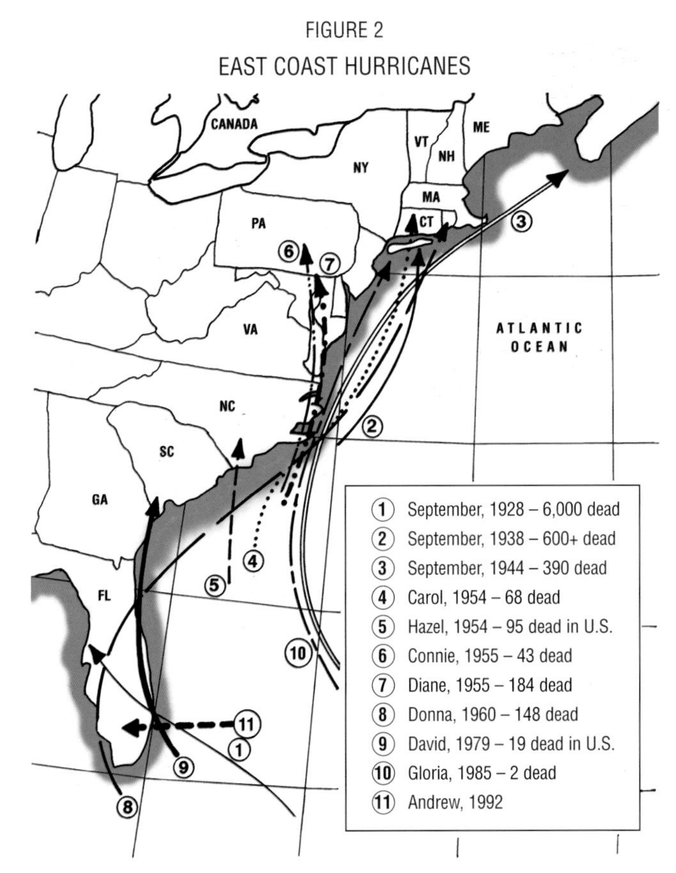

Hurricanes before WWII were not named. Source: National Oceanic and Atmospheric Administration

TABLE 3

MEAN ANNUAL NUMBER OF THUNDERSTORMS

Hartford	26
New Haven	27
Norfolk	32
Storrs	19

Source: Brombach, Climate of Connecticut.

Duration is the length of time over which a precipitation event lasts. Typical events range from thunderstorms with a duration of less than an hour to low-pressure systems of several days.

Intensity is the depth of rainfall per time period, usually measured in inches per hour. People who call a rainstorm "light" or "heavy" are describing its intensity. Highest rain intensities generally occur for short durations, while the average intensity of long-duration events is lower.

Depth and intensity of precipitation are key factors in planning storm drainage systems, such as pipes, channels and dams, which must have the capacity to handle the maximum rainfall amounts and intensity that occur during rare weather events.

During colder months, from November to April, the primary source of precipitation in Connecticut is from low-pressure systems that originate in the Gulf of Mexico, along the East Coast or in the Southern states. These "northeasters" (so-called for the prevailing direction of the initial winds) approach New England from the south, drawing moisture from the ocean and depositing long-duration rain or snow along the coast and in the southern hills of the eastern and western highlands. Connecticut receives about 20 of these storms each year.

Storms of greater intensity usually occur in the state during warm weather, from June to September.

Shorter-duration events with the highest intensity — the kind that can cause flash floods — are almost always summer thunderstorms. These storms form in warm, unstable air masses, usually moving from west to east. They are most frequent in July, though they can occur during any month. Thunderstorm activity is generally greatest in the Litchfield Hills and parts of southwestern Connecticut, where the terrain is rugged, and less frequent in the southeast, where the ocean air has a moderating effect. (See Table 3.) Thunderstorms occur on from 18 to 35 days per year in Connecticut.

Occasionally, Connecticut is subjected to longer-duration storms of high intensity — those producing record rainfall depths — known as tropical storms and hurricanes. The latter produce winds of at least 75 miles per hour and frequently torrential rain. Coastal sections of Connecticut have slightly more intense, long-duration rainfall than inland sections because of such severe maritime storms.

Tropical storms may occur from early summer through mid-autumn, though they are most common in August and September. Hurricanes generally have set the records for intense long-duration precipitation.

The location of southern New England in relation to the frequent track of hurricanes coming up the East Coast, as shown in Figure 2, makes the region particularly vulnerable to these intense storms.

Fortunately, the forces that produce the heavy rain usually tend to move tropical storms rapidly along the coast, so that no area is subjected to heavy downpours for a long period. Rarely, however, unusual circumstances combine to produce extreme rainfall amounts. In 1955, for example, Hurricane Diane struck Connecticut, dropping 15.92 inches of rain from August 17 through 19. During one 24-hour period alone, rainfall totaled 12.77 inches.

Streamflow

The rate at which water flows in rivers and brooks is called streamflow. It varies by region, and within a given region can fluctuate widely depending on season, weather, topography, and other factors (see Table 5, Page 14).

Streamflow is measured as a quantity of water per unit of time. Instantaneous streamflow, also called the stream's discharge rate, is usually expressed as the volume of water, in cubic feet per second, passing a given point. Water companies often measure flow in terms of millions of gallons per day. Long-term runoff can also be expressed in the number of inches of equivalent precipitation.

Probably the most obvious variation in streamflow is from season to season. Within a given year, most rivers have distinct periods of high flow followed by periods of low flow. The former results from storm runoff and snowmelt during late winter and spring, the latter from the variable precipitation and high evapotranspiration rates common during summer and fall. The variation in a river's flow rate over time can be expressed in a diagram known as a hydrograph.

Annual runoff is the net result of all of the natural influences, as well as the effects of human activities. Whatever the natural variations between geographic regions, each has a characteristic pattern called the streamflow regimen. (See Figure 3.)

In Connecticut, total annual streamflow ranges from the equivalent of about 22 inches of precipitation per year in the north central and southwestern parts of the state to 29 inches in the southeast. Flow rates are measured at gauging stations throughout the state. Runoff extremes and average discharges of several representative streams are shown in Table 4.

FIGURE 3

SEASONAL PATTERN OF STREAMFLOW IN CONNECTICUT

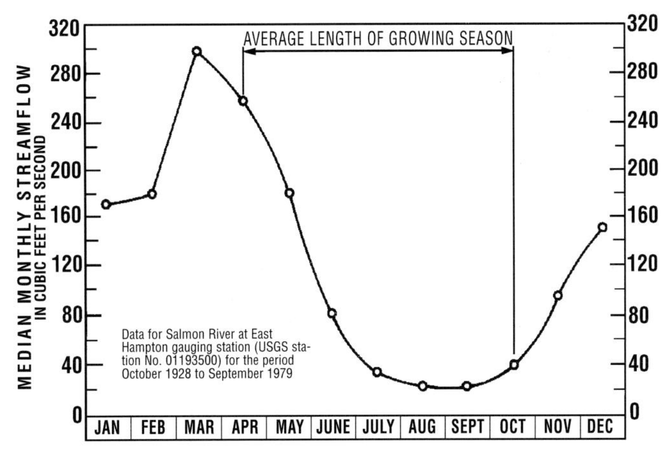

Source: United States Geological Survey

This streamflow regimen illustrates the typical seasonal pattern of flows. During the growing season, most precipitation is returned to the atmosphere by evaporation and transpiration and there is relatively little surface runoff or groundwater recharge.

TABLE 4

NATIONAL WATER SUMMARY — SURFACE-WATER RESOURCES

SELECTED STREAMFLOW CHARACTERISTICS OF PRINCIPAL RIVER BASINS IN CONNECTICUT

GAUGING STATION			STREAMFLOW	
STATION NAME	DRAINAGE AREA (square miles)	PERIOD OF ANALYSIS	7-DAY, 10-YEAR LOW FLOW (cubic feet per second)	AVERAGE DISCHARGE (cubic feet per second)
NEW ENGLAND REGION—Connecticut Subregion—Connecticut River basin				
Connecticut River, at Thompsonville	9,661.0	1928-83	2,200.0	16,400.0
Burlington Brook, near Burlington	4.10	1931-83	0.7	8.3
Farmington River, at Rainbow Dam	590.0	1928-60	144.0	1,030.0
		1961-83	101.0	1,040.0
Salmon River, near East Hampton	100.0	1928-83	5.2	184.0
CONNECTICUT COASTAL SUBREGION—Thames River basin				
Mount Hope River, near Warrenville	28.6	1940-83	0.9	51.2
Shetucket River, near Willimantic	404.0	1926-52	46.5	667.0
		1953-83	44.2	734.0
Quinebaug River, at Jewett City	713.0	1916-58	119.0	1,250.0
		1959-83	90.0	1,330.0
Yantic River, at Yantic	89.3	1930-83	5.2	165.0
QUINNIPIAC RIVER BASIN				
Quinnipiac River, at Wallingford	115.0	1930-63	32.6	211.0
HOUSATONIC RIVER BASIN				
Housatonic River, at Falls Village	634.0	1912-83	119.0	1,090.0
Shepaug River, near Roxbury	132.0	1930-71	6.2	236.0
Pomperaug River, near Southbury	75.1	1932-83	6.0	128.0
Housatonic River, at Stevenson Dam	1,544.0	1926-83	160.0	2,600.0
Naugatuck River, at Beacon Falls	260.0	1926-59	61.2	484.0
		1960-83	59.4	557.0
SAUGATUCK RIVER BASIN				
Saugatuck River, near Westport	79.8	1932-67	2.25	119.0

Gauging station: Period of analysis is for the water years used to compute average discharge and may differ from that used to compute other streamflow characteristics.

TABLE 4

Streamflow characteristics: The 7-day, 10-year low flow is a discharge statistic: The lowest mean discharge during 7 consecutive days of a year will be equal to or less than this value, on the average, once every 10 years. The average discharge is the arithmetic average of annual average discharges during the period of analysis.

100-year flood: That flow that has a 1% chance of being equaled or exceeded in a given year.

CHARACTERISTICS

100-YEAR FLOOD (cubic feet per second)	DEGREE OF REGULATION	REMARKS
209,000	Appreciable	Regulation by power plants in Vermont and Massachusetts. Diversion for water supply in Massachusetts. Flood control for approximately 20 percent of basin. National stream quality accounting station.
1,250	Negligible	Index station for long-term trends in natural streams.
44,000	Appreciable	Regulation by power plant at Rainbow Dam. Diversion for water supply of Hartford-area Metropolitan District Commission. Flood control since April 1960.
24,000		
16,600	Negligible	Index station for long-term trends in natural streams.
5,620	Negligible	Index station for long-term trends in natural streams.
25,000	Moderate	Flood control since March 1952. Diversion for water supply of city of Willimantic.
22,500		
29,500	Moderate	Flood control since September 1958. NASQAN station.
26,500		
10,800	Negligible	
6,340	Moderate	Diversion for water supply of city of New Britain.
24,000	Moderate	Regulation by Falls Village power plant.
24,000	Appreciable	Diversion for water supply of city of Waterbury.
19,900	Negligible	Minor diversion for water supply of town of Woodbury. Index station for long-term trends in natural streams.
95,100	Appreciable	Regulation by power plant at Stevenson Dam. Diversion for water supply for city of Waterbury. Some flood control. NASQAN station.
46,000	Appreciable	Flood control since December 1960. Diversion for water supply of city of Waterbury.
23,000		
13,400	Appreciable	Diversion for water supply of several communities.

Sources: Reports of the U.S. Geological Survey and State of Connecticut Department of Environmental Protection

TABLE 5

TYPICAL STREAMFLOWS IN CONNECTICUT

RUNOFF EVENT	CU. FT. PER SECOND PER SQ. MILE OF WATERSHED (cfs/sm)	COMMENTS
7-day duration low flow	0.02-0.2	Varies with stratified drift area for 10-yr. frequency (7Q10)
August mean	0.2-0.6	Varies with stratified drift area
Annual median	1.0	
Annual mean	1.8	
March mean	2.5-5.0	
Mean annual flood	20-40	For 100-sq.-mile watershed
	30-50	For 10-sq.-mile watershed
	40-60	For 1-sq.-mile watershed
100-year flood	3-10 times mean annual flood	Varies with location, watershed soils, slope, flood storage capacity, and land use

RARE FLOOD FLOWS

"Typical" rates	200-500 (cfs/sm)	August 1955 flood
Extreme rates	500-1,000	August 1955 flood
"Typical" rates	200-400	June 1982 flood
Extreme rates	650	June 1982 flood

A spring is a groundwater discharge emanating from a well-defined point.

Precipitation can cause rapid changes in streamflow. Flow increases initially as rain falls directly into rivers, lakes, and ponds, and as it runs off nearby paved surfaces. Streamflow continues to rise if rainfall is heavy enough to cause runoff from saturated soils or if the rainfall rate exceeds the soils' infiltration rate. Brooks and rivers will have higher peak flows in steep and/or relatively impermeable watersheds, where runoff flows rapidly downhill, allowing less opportunity for it to evaporate or infiltrate into the ground.

Streamflow begins to decline when precipitation ends and runoff decreases. The rate of flow decline depends largely on how long it takes water from tributary streams, overland flow areas and interflow discharges to reach the main channel.

Topography and climate can affect streamflow in other ways. Precipitation usually is greater at high elevations. Also, storms moving across a mountain range, from west to east for example, drop most of their precipitation on the western slopes, contributing more moisture to that watershed than the one on the eastern slope, which is said to be in the "rain shadow." In mountainous areas and other regions with cold climates, streamflows are more independent of the monthly distribution of precipitation, because low temperatures cause the storage of precipitation as snow or ice, delaying runoff.

Connecticut provides an excellent example of the effect cold has on runoff rates. In southern Connecticut, snow melts relatively quickly and has limited effect on seasonal variations in runoff. Where snow is stored in only part of a basin, or if it melts occasionally during the winter — as in much of interior Connecticut — the snow storage effect is only slightly greater. In contrast, the Connecticut River usually has a prolonged spring flood as the snowpack in northern New England melts and is released into the river and its tributaries.

Between precipitation events, a river's base flow — the amount that continues to flow even during prolonged dry weather — depends largely on the discharge of groundwater from springs and direct discharges into the channel, and on the drainage from wetlands and lakes, many of which store water during periods of high flow, releasing it gradually during periods of lower flow.

Groundwater Discharges

When a stream's water level is lower than the water table (Figure 1), groundwater seeps into the channel, replenishing the flow. This storage effect is so great in some areas that the groundwater it releases may constitute 40 percent of the annual streamflow, and nearly all of the flow during dry seasons.

Where the water table intersects the surface of the land, water emerges as springs and seeps. These form the headwaters of many small streams and wetlands, and are important to sustaining streamflow during dry weather.

A spring has a concentrated flow, while a seep is a poorly defined discharge emanating from a larger area of soil or rock rather than from one point. Either may occur in a variety of geologic settings: bedrock fractures, confined groundwater that creates artesian springs, water perched over rock, till, or impermeable soil, or thick, permeable aquifers. Streams associated with the latter type of aquifer tend to have higher base flows with less annual variation.

Water discharged by springs and seeps is usually cool, clean, and rich in dissolved oxygen, which is crucial for aquatic life. Water emerging from shallow aquifers has a temperature similar to the region's mean annual temperature: It feels cool in the summer, but mild in the winter in comparison to the surface waters.

The destructive power of a flooded river.

Building in floodplains has almost inevitable costs.

Floods

A flood is a high flow that overtops a stream's natural channel or artificial confines — a natural event in watersheds. Floods can range from a slight rise above the stream's banks to a raging torrent that may inundate a wide area.

Floods can have beneficial effects, scouring debris from the stream channel, allowing migratory fish to pass over rapids, and depositing rich sediments on floodplains, but they can cause extensive damage to inhabited areas.

Key factors affecting a flood's intensity include rainfall patterns, watershed characteristics, channel capacity, and land use.

Precipitation within a watershed, especially the intensity and duration of rainfall, is the principal factor influencing stormwater runoff. In small watersheds, the relationship between rainfall and runoff is much more direct than in large watersheds, where channel storage, snowmelt, and base flow have greater influence.

Attributes of a watershed that affect runoff timing include its till content, shape, length, slope, and friction roughness. For instance, the surface runoff and interflow in steep watersheds travel faster than in flat watersheds and this tends to concentrate the runoff into a shorter time span with a higher peak flow. Similarly, in short, wide watersheds, the

High water provides an exciting challenge to experienced kayakers.

runoff does not have to travel very far and so is concentrated more rapidly in the channel than it is in long, narrow watersheds. Watersheds with smooth ground surfaces, impervious surfaces, man-made channels, and underground pipes will have higher runoff velocities and shorter concentration times than will heavily vegetated watersheds.

Wetlands and floodplains help to reduce the peak flood flow rates and crest elevations. Such areas temporarily store a portion of the excess runoff and delay its movement downstream.

Intense rains, rapid snowmelt, or a combination of the two can create havoc. Larger floods are often caused by intense rainfall in more than one tributary basin. The peak runoff from different tributaries may arrive all at once at their confluence, creating very high water levels.

Fortunately, in Connecticut, snowmelt generally has a minimal effect on annual peak flows, because the snow tends to melt gradually, especially in rural areas. Snow accumulation in those areas rarely exceeds 2 feet in depth, and rainfall rarely exceeds 1 inch when snow is on the ground. An exception is in the Litchfield Hills, where snowpacks are much greater than elsewhere in the state and snowmelt has a greater effect on peak streamflows.

Major flooding can occur at any time of year in Connecticut.

During winter and early spring, snowmelt can combine with rainfall, as happened in the flood of March 1936. Winter and spring flooding can be exacerbated by ice jams, such as those that occur periodically on the Housatonic River in New Milford and the Salmon River in East Haddam.

In the summer, severe thunderstorms often cause flash flooding in small watersheds, and more widespread flooding can result from tropical storms, as occurred in southern Connecticut in June 1982.

During late summer and fall, hurricanes can be the pivotal factor, causing floods such as those in September 1938 and August and October 1955.

Floods can be measured in various ways: according to the probability that a flood of a certain intensity will occur in a given year, by the height of their crests, or by their peak flow rates.

The first method describes rare events in terms of the likelihood that the flood flow will be equaled or exceeded in any one-year period. For example, if there is a 1 percent chance that a flow will be equaled or exceeded in a given year, statistically one such event will occur every 100 years on average. Such an event is commonly termed a 100-year flood.

The extent of inundated land can be determined from historical evidence, such as high water marks or observations of residents. Or it can be estimated by indirect methods that may include surveying the channel and mathematically computing the impact of various flows.

Flood flow rates and probabilities of recurrence on large rivers are determined from gauge records. Flood magnitude may also be estimated from the amount of runoff likely to be produced by a storm of specified frequency or from the most intense foreseeable storm.

The comparison of normal streamflow rates with records from two floods illustrates the extremely high runoff rates that intense, rare storms can produce.

Connecticut's long-term average runoff rate ranges from 1.6 to 2.0 cubic feet per second for each square mile of watershed. During the August 1955 flood, many rivers had peak flows ranging from 200 to 500 cubic feet per second per square mile of watershed, with a few smaller rivers — whose watersheds were less than 20 square miles — reaching peak flows of almost 1,000 cubic feet per second per square mile. Peak flow rates during the June 1982 flood were generally 200 to 400 cubic feet per second per square mile, with two rivers near the storm center recording rates of 644 and 660.

Traditionally, in response to flood threats, people have built dams to temporarily detain water in flood retention areas, thus reducing peak flows, or have constructed large artificial channels to convey them harmlessly away from developed areas. Since the 1955 flood, the U.S. Army Corps of Engineers and the Natural Resource Conservation Service (formerly the U.S. Soil Conservation Service) have built reservoirs in the Quinebaug, Farmington, Park, and Naugatuck river basins, and have enlarged many miles of river channels.

Today, the emphasis has shifted to also include land-use planning to minimize the increase in runoff that results from impervious surfaces, to limit channel and floodplain encroachments, to conserve wetlands, and to protect riverine ecosystems from erosion and water-borne pollutants.

Construction of dams to regulate streamflow has been a traditional means of reducing flood damage in river valleys.

Droughts

A drought, in general terms, is a dry period severe enough to affect natural vegetation, crops, and human activities. The severity of a drought depends on its duration, rain deficit, geographical extent, and availability of alternate sources of water.

Short-term drought conditions in Connecticut generally occur in

Man-made reservoirs store water for use during periods of dry weather.

summer and early fall; the exceptionally dry summer of 1995 is a good example.

Connecticut's most notable long-term drought of the 20th century occurred in the mid-1960s, during which there were serious water-supply concerns.

Lack of rain for a few weeks or a month may have no appreciable effect, except for the depletion of small reservoirs if water demand is high. As a drought continues, however, usable water in storage, both surface and underground, is progressively depleted. Rivers have less water available for drinking, waste assimilation, fisheries, and recreation. In severe cases, salt water intrudes from the coast.

Measures that can reduce the effect of drought on communities include having adequate water-supply storage reservoirs, water conservation, and proper maintenance of water-supply systems, including control of pipe leakage.

The effect a drought has on streamflow varies with a number of factors, including rainfall deficit, watershed area and watershed geology. Since a watershed's base flow between rainstorms and periods of snowmelt is related to groundwater discharge, a watershed underlain by permeable stratified drift that can store water and release it during dry spells will have higher flows during droughts than will watersheds underlain mostly by bedrock or impermeable till.

In fact, a watershed's base flow has been found to vary according to its percentage of stratified drift, which has a high capacity to store water between particles.

The most widely used index of a stream's low flow during droughts is the lowest average flow for seven consecutive days that occurs with a 10-year frequency (called by engineers the 7Q10 flow).

Watersheds of less than five square miles with little stratified drift deposits may have no runoff at all during a 7Q10 event. Even watersheds twice that area may have streamflows of only 0.1 cubic foot per second for each square mile if they are mostly impermeable.

Larger watersheds with 40 percent stratified drift will have runoff of about 0.3 cubic feet per second per square mile during such extreme conditions.

Because these low base flows are dependent on groundwater discharges, during droughts the water quality in streams where waste assimilation and pollution are not a concern is similar to the groundwater quality, with elevated levels of dissolved minerals and low concentrations of suspended sediments.

HYDRAULICS — THE FLOW OF WATER

A scientific understanding of the flow and other characteristics of surface water is crucial to the implementation of appropriate flood-control measures, development of adequate water-supply systems and adoption of wise land-use regulations.

Hydraulic engineers have many methods of classifying and analyzing the flow of water in open channels — whether natural or artificial — as gravity draws it down toward the sea. (See Table 6.)

Natural channels such as rivers, brooks and streamlets usually have varying dimensions, slopes, and alignments. Thus, their water flow is usually non-uniform.

Canals, ditches, gutters, pipes, and other artificial, man-made channels usually have constant dimensions and so allow the possibility for uniform flow.

Most analyses of rivers involve the study of steady flow rates that are constant for a specified condition. Occasionally, unsteady flow is encountered during flash floods and at stormwater detention basins, reservoirs, and tidal channels, where the flow rates vary rapidly with time.

The simplest and most common methods of computing and predicting steady flow conditions — including velocity, depth and rate — are valid only for uniform flow. These analytical equations take into account the channel size, shape, and friction.

The use of more complex techniques is usually required to approximate non-uniform flow when river channels vary significantly in size or slope, when a stream is flowing both in a channel and on a floodplain, and for flow at bridges, dams and culverts. The equations used for studying non-uniform flow are based upon the conservation of energy.

The analysis of unsteady flow is complicated because a river's discharge rate varies with time. For example, to predict the future water level in a reservoir, one has to keep track of all of the water entering and leaving the reservoir each day, including diversions, leakage, rainfall and evaporation. The volume of water stored in the reservoir would be equal to the initial water volume, plus the sum of the daily inflow, minus the daily outflow.

The analysis of detention basins, tidal rivers, and floodplains with large storage areas also involves unsteady flow and requires advanced hydraulic principles such as conservation of mass and momentum in order to evaluate flow conditions.

TABLE 6

TYPES OF WATER FLOW

Steady flow	The movement of water at a constant rate.
Unsteady flow	A rate of flow that changes rapidly, such as when runoff is increasing due to rainfall or when streamflow is affected by ocean tides.
Uniform flow	A condition where the water depth is constant along a length of channel and the water velocity is constant at all points of a cross-section. This requires that the channel slope, roughness, and size also be constant, such as in a man-made channel.
Non-uniform flow	Flow with variable depths, widths, and slopes along a channel. This is a common condition in natural river channels.
Laminar flow	Occurs when the water droplets in a stream move in smooth parallel paths, influenced by the fluid's viscosity.
Turbulent flow	In contrast to laminar flow, water particles move in irregular, converging or diverging paths. This is the normal situation in natural rivers.
One-dimensional flow	Water particles move in one direction on parallel paths, without lateral flow. May be either laminar or turbulent.
Two-dimensional flow	Water particles move in any horizontal direction, such as the combination of downstream channel flow and lateral floodplain flow around obstructions.

TABLE 7

USE OF FRESH WATER IN CONNECTICUT IN 1990

Water use in millions of gallons per day

OFFSTREAM WATER USE:	SURFACE	GROUND	TOTAL
Public Water Supply	301	73	374
Self-supply:			
Agricultural (irrigation)	6.5	8.2	15
Agricultural (livestock)	0.4	1.1	1.5
Commercial	0.6	18	18
On-site private residential	0	46	46
Industrial	61	19	80
Mining	1.8	0.4	2.2
Thermoelectric Power	530	0.2	530
1990 State Total	902	165	1,070
INSTREAM WATER USE:			
Hydroelectric Power Generation	4,152	0	4,152

"Estimated Water Use in the United States in 1990, U.S. Geological Survey Circular 1081, 1993

GROUNDWATER WITHDRAWALS IN 1985

TYPES OF AQUIFERS	MGD	PERCENT	PRIMARY USE OF AQUIFER
Stratified-drift	84	58%	Public Water Supply
Crystalline bedrock	45	31%	Private Wells[1]
Sedimentary bedrock	11	8%	Private Wells[1]
Carbonate bedrock	4	3%	Private Wells[1]
1985 State Total	144	100%	

[1] Combined commercial and domestic

Source: Connecticut Water-Use Information Program • Natural Resources Center, DEP

The size of a watershed has significant effects on the water flow patterns within it.

Watersheds considered small from a hydrologic point of view — from a few acres to 100 square miles in area — are dominated by the effect of overland flow, respond rapidly to short-duration, high-intensity rainfall, and are quite sensitive to changes in land use. Their waters typically flow over the surface and in small channels such as rills, gullies, and channels of the first, second, and third order (see Chapter Two).

Large hydrologic watersheds have long flow travel times (more than 24 hours to traverse the watershed's length, for example) and have peak flow patterns that are modified by channel storage and floodplains. Because of those characteristics and the time it takes runoff to reach the main channel, peak flow rates in large watersheds are not very sensitive to intense, short-duration rainfall and peak flows may occur long after precipitation has ended.

WATER: ESSENTIAL FOR LIFE

Water is inextricably linked to human life. Not only is it necessary as drinking water and to grow the food that nourishes us, but it also is used to generate the electricity used in our homes and industries and to produce the many durable goods society requires. We also use it to dispose of many of the industrial, commercial, and domestic wastes we create. (See Table 7.)

Water use can be classed in two broad categories: instream and offstream. Instream uses are those that do not require the withdrawal of water from a lake or stream channel. They include generation of hydroelectric power, navigation, recreation such as fishing and boating, and use as wildlife habitat. Offstream uses involve withdrawals of water, for irrigation, drinking, industry, etc.

In 1990, average offstream water use in Connecticut totaled 325 gallons per person per day, well below the national average. The average per capita use from public water supplies was 140 gallons per day.

Connecticut uses a total of more than 5 billion gallons of water per day. However, most major users in Connecticut do not "consume" water. Water is often withdrawn from rivers and quickly returned. Thus, it is available for reuse.

For example, treated wastewater may be discharged to a river and then circulated through the condenser of a downstream thermoelectric power plant. The water may further be used in a non-consumptive manner to drive the turbine of a hydroelectric plant or in a consumptive manner to irrigate a farmer's field.

Surface water is an invaluable natural resource. About 4 billion gallons per day is used in Connecticut to generate hydroelectric power. Another 400 million gallons a day is withdrawn by public water supplies for use by households and businesses. The state's extensive development and the increasing demands on its water resources, including the dependence of urban areas on reliable supplies of water, has led to increased regulatory control over flow diversions.

Today, the water quality of streams in Connecticut makes them suitable for most intended uses, largely due to a pollution control program instituted in 1967. For instance, Connecticut is one of only two states in the country that prohibit wastewater discharges in the watershed of a public water supply reservoir. The pressing issue for the future is how to maintain high quality in the face of increasing demands.

WATER SUPPLY RESERVOIRS

Reservoirs are the primary source of potable water for Connecticut's metropolitan areas.

Approximately 84 percent of the state's population is served by public water systems, and about 80 percent of these public supplies are from surface water sources, either reservoirs or river intakes. There are over 200 public water supply reservoirs in the state, of which most are currently active (See Table 8).

Most water supply reservoirs in Connecticut are formed by damming rivers to create large impoundments. These reservoirs store water from periods of high runoff for use throughout the year, especially during summer low flow periods.

The long-term water-supply yield of reservoirs varies with watershed size, the seasonal patterns and volume of runoff, and reservoir storage volume. In New England, reservoir yields typically vary from about 600,000 gallons per day per square mile of watershed to 1 million gallons per day.

Although the primary purpose of water supply reservoirs is to meet their regions' current and projected demands for water, reservoirs with excess capacity often simultaneously provide for many other uses. These include flood management and regulating downstream river flow for waste assimilation, recreation, aquatic habitat,

TABLE 8

PUBLIC WATER SUPPLY AND PRIVATE SELF-SUPPLY WATER USE IN CONNECTICUT IN 1985

	Population Supplied	MGD†
SURFACE WATER		
Public Supply	2,165,050	296
Private Wells	0	0
	2,165,050	296
GROUND WATER		
Public Supply	518,190	66
Private Wells	514,990	39
	1,033,180	105
SURFACE & GROUND		
Public Supply	2,683,240	362
Private Wells	514,990	39
	3,198,230	401

POPULATIONS SERVED:

2,683,240 people	(84%)	on public water supply
514,990 people	(16%)	on private domestic wells
3,198,230		
2,165,050 people	(68%)	dependent upon surface water
1,033,180 people	(32%)	dependent upon ground water
3,198,230		

WATER QUANTITIES WITHDRAWN:

296 MGD	(75%)	from surface water (reservoirs)
105 MGD	(26%)	from ground water (wells)

NUMBER OF WELLS USED:

3,300 non-community wells serving transitory populations

1,500 public water supply wells serving	518,190 people
250,000 private domestic wells serving	514,990 people
	1,033,180

†MGD = Million gallons per day
*Domestic = Domestic Self-Supply (private backyard domestic wells)

Connecticut Water-Use Information Program Natural Resources Center, DEP

FIGURE 4

MAJOR DRAINAGE BASINS IN CONNECTICUT

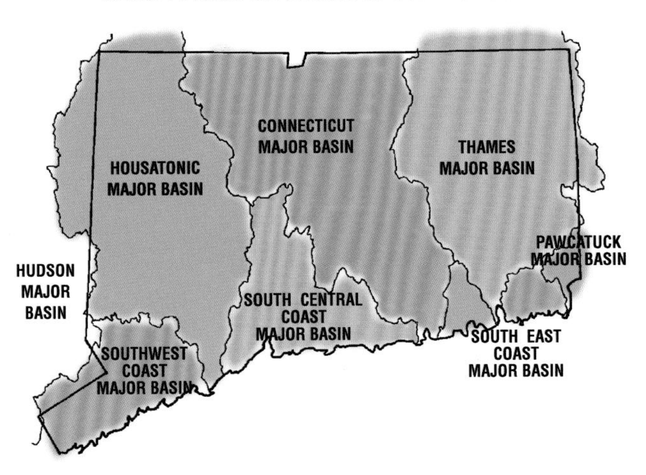

TABLE 9

A SUMMARY OF SURFACE WATER RESOURCES IN CONNECTICUT

Number of major water basins according to state subdivisions	8
Total number of river miles	8,400 miles
Number of river border miles	11 miles
Number of lakes/ponds/reservoirs	6,500
Area of lakes/ponds/reservoirs	82,900 acres
Area of estuaries/harbors	600 square miles
Miles of coastline	253
Area of freshwater wetlands (15 percent of state area)	435,000 acres
Area of tidal wetlands	17,500 acres
Names of border rivers	Byram River (one mile, New York) Pawcatuck River (10 miles, Rhode Island)

hydropower, and navigation.

Since many of these uses compete for a limited supply of water, the design and construction of dams require careful planning in order to meet water supply needs while minimizing impact on the riverine environment.

CONNECTICUT'S WATER RESOURCES

Connecticut has an extensive network of surface water resources, including rivers and streams, ponds and lakes, fresh and brackish wetlands, and coastal estuaries. These riches are the result of the state's humid climate, its low-permeability uplands, and the rolling countryside with its many small watersheds.

If the state's approximately 8,400 miles of rivers and smaller streams were added together end to end, they would reach one-third of the way around the Earth.

Connecticut also has about 6,000 lakes, ponds, and reservoirs, covering a total of 82,900 acres. Most are man-made, resulting from the construction of dams across streams and rivers. Other ponds were created by excavating holes deep enough to extend below the water table.

All of Connecticut's land drains into rivers that discharge into Long Island Sound, a semi-enclosed estuary where the freshwater runoff mixes with saline ocean waters. Consequently, the health of Long Island Sound is strongly influenced by the quality of the water in Connecticut's rivers.

The Connecticut Department of Environmental Protection has identified eight major drainage basins in the state (see Figure 4). These major basins are divided into smaller areas classified as regional basins, each representing the watershed of a mid-size river; subregional basins, generally the drainage areas of tributaries to mid-size rivers; and local drainage basins, the smallest category.

The Thames major drainage basin, for example, which comprises much of eastern Connecticut, is divided into nine regional basins — corresponding to the watersheds of the Willimantic, Natchaug, French, Fivemile, Moosup, Pachaug, Quinebaug, Shetucket, and Yantic rivers — which in turn are made up of 76 subregional basins.

Channels and Floodplains

> ***Let us permit nature*** to have her way: she understands her business better than we do.
>
> Michel de Montaigne (1533-1592), *Essays*

TABLE 10

CHANNEL CHARACTERISTICS

CHANNEL ALIGNMENT	CHANNEL TYPE	TYPICAL ENVIRONMENT	COMMENTS	SLOPE
(a) (b) 	(a) Regular serpentine meanders (b) Regular sinuous meanders	Ancient glacial lake beds, very mild slopes with free bends	Uniform cohesive materials; low, steep banks; pools and riffles	Low
	Tortuous or contorted meanders, no cutoffs	Misfit stream in glacial soils	Uniform cohesive materials	Low
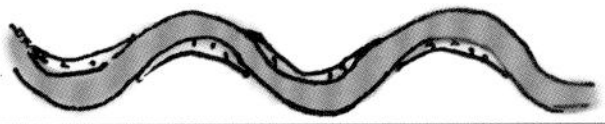	Meanders with point bars	Glacial outwash, sand-filled mild to moderate slope channel	Slightly cohesive top stratum over sands and gravel; pools and riffles	Low to moderate
	Unconfined random meanders with oxbows, scrolled	Sandy to silty deltas and old alluvial floodplains	Slightly cohesive top stratum over sands or fine-grain sediments	Very low
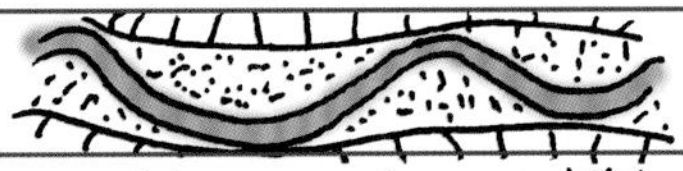	Confined meandering, forced bends	Steep-walled channel	Slightly cohesive top stratum over sands	Low to moderate
	Deep entrenched meanders with fixed bends	Glacial till or erodible bedrock	Till, boulders, soft rock; steep banks	Moderate
	Meanders within meanders	Misfit streams in large glacial valleys	Cohesive materials	Low
	Irregularly sinuous	Thin till over bedrock in plains	Glacial till, low sediment load, rapids and pools	Moderate to steep
	Irregular non-uniform	Foothills and mountain valleys	Cobble-veneered sand	Moderate to steep
	Anastomosing, multiple channels	Foothills, plains	Sand-bed or gravel-paved channel, high sediment loads	Low
	Braided	Glacial outwash; foothills	Sand and gravel, many mid-channel bars, high sediment loads	Medium
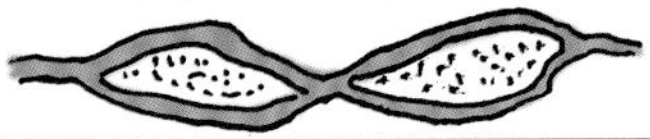	Irregular channel splitting	Large rivers in bedrock or with many boulders	Alternate sand, gravel and rock	Varies
	Rectangular channel pattern	Jointed rocks, mostly flat-lying sedimentary rocks	Shallow sedimentary bedrock	Varies
 L R L	Lakes (L) and rapids (R)	Glacial till, shallow bedrock	Till cobbles, boulders; hard rock	Irregular
	Straight, incised	Mountain valleys, bedrock, steep bed slope, ravines	Till or bedrock fault lines, waterfalls and cascades	Steep
	Straight, gully	Eroded into till or stratified drift, steep slopes.	Sand and gravel, high erosion rate; step-pod profile	Steep

Channels and Floodplains

Watercourses range from small, clear, headwater brooks to large rivers full of sediment.

This section describes the dynamic geologic and hydraulic processes associated with their origin, formation, growth, and flow toward the sea. The types of channels discussed in this chapter are shown in Table 10.

HOW CHANNELS FORM

When the amount of precipitation is greater than the soil can hold and it does not evaporate, water runoff flows downhill across the ground's surface. As it follows the path of least resistance, this thin layer, called sheetflow, is interrupted by trees, rocks, litter, and other objects, concentrating it into paths and forming rivulets. Moving down the slope, these rivulets grow in width and depth as they receive more runoff from nearby areas.

In the larger rivulets, the water dislodges the leaves and twigs that cover the ground and sweeps them downhill. It also erodes loose soil, forming depressed channels, which grow wider and deeper as more and more rivulets enter them. Convergence of these larger streamlets eventually forms well-defined watercourses.

A swimming hole along a bedrock channel.

STREAM ORDER

The size of watercourses and their relative positions within a watershed are described by a system known as stream order, which defines the sequence in which small streams flow into larger ones and the hierarchy of the various tributaries of larger rivers. (See Figure 5, and Table 10A.)

A first-order stream is a perennial stream so small that it does not have any mappable tributaries. (U.S. Geologic Survey topography maps, at a scale of one to 24,000, are used as the standard reference. They represent first-order streams as solid blue lines, disregarding rivulets, rills, or intermittent channels.)

TABLE 10A

Stream Order	Typical Width in Feet
1	0-5
2	5-10
3	10-25
4	25-100
5	100-250
6	Over 250

FIGURE 5
STREAM ORDER SYSTEM

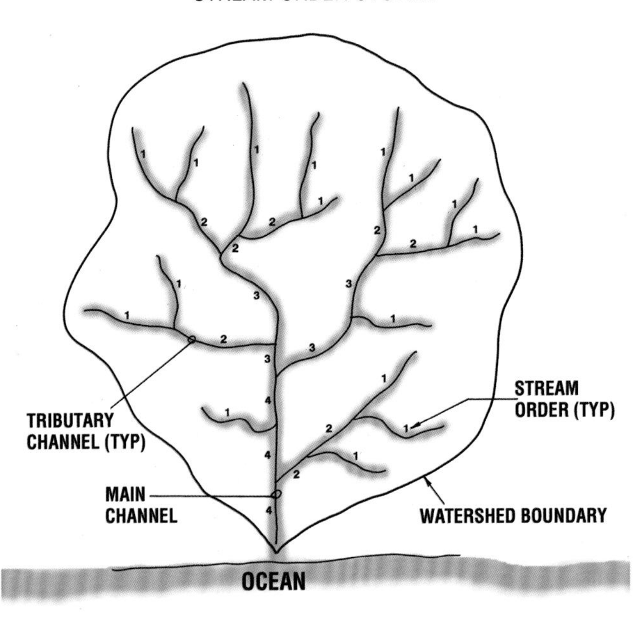

TABLE 11
STREAM ORDERS OF SELECTED CONNECTICUT RIVERS

RIVER	STREAM ORDER
Connecticut River	6
Thames River	6
Housatonic River	5
Quinebaug River	5
Farmington River	5
Naugatuck River	4
Quinnipiac River	4
Hockanum River	4
Salmon River	4
Shepaug River	4
Blackledge River	3
Hop Brook	3
Nonnewaug River	3
Broad Brook	3
Fawn Brook	2
Exeter Brook	2
Transylvania Brook	2

Second-order streams have only first-order streams as their tributaries. A third-order stream can have first- or second-order tributaries. A higher order is created only when two watercourses of the next-lowest order join. For example, the channel that results from the junction of two first-order streams is a second-order stream. The third order is not attained until that stream converges with another of the second order. The junction of two third-order streams creates a fourth-order stream, and so on.

Typically, first-order streams are less than a mile long, with small watersheds, narrow channels, and limited flow rates. They seldom support large fish and are too small for most recreational activities.The United States Fish and Wildlife Service estimates that such streams represent 40 percent of the total length of perennial streams nationwide.

First-, second-, and third-order streams are considered to be headwaters; their principal function is to collect runoff. Watercourses of orders four and five are intermediate in size. They are usually large enough to have floodplains and to provide for many forms of recreation. The larger orders, six and above, are of great significance. Often these rivers are navigable.

The Connecticut Department of Environmental Protection has identified stream orders as part of its delineation of river basins. The stream order of selected rivers at their mouths is shown in Table 11.

HOW RIVERS DEVELOP

According to classical geologic theory, streams mature through three stages.

A youthful stream is characterized by a steep, narrow channel and lack of a floodplain. (See Figure 7.) Its profile — a term used to describe the channel's slope — is irregular, with waterfalls, rapids, and lakes that are gradually reduced through erosion. Such channels receive limited sediment from upstream. They tend to have clear, cold water rich in oxygen. When vertical erosion is faster than horizontal erosion, deep V-shaped valleys are formed. Such streams are seldom navigable, and the narrow valleys allow little room for most human activities. However, such valleys have often been dammed, with the resulting impoundments used to power mills or hydroelectric plants.

Streams are considered to be mature when their channel beds have smooth, uniform profiles, free of abrupt grade changes. Portions of banks may be eroded by floods, only to be refilled by sediment at other times.

In a mature stage, a stream's downcutting will slow or even stop, due to reduced slope and velocity. However, lateral erosion and deposition continue, forming a valley broader than the river, as the latter meanders from side to side along its course. Development of a small floodplain on the valley floor is a sign of an early stage of maturity. Such channels often occur in stream orders four through six.

As a river enters old age, the width of the active meander belt increases. By now, its floodplain is broad. The meander belt may eventually be 10 to 20 times wider than the river itself, with the floodplain being significantly broader than that. The floodplain may be scarred by the river's wandering channel, with oxbow lakes, clay plugs, natural levees, and backwater swamps.

Such established floodplains have long been used for human activities. The generally level land is easy to develop, and settlement has been encouraged by trade made possible by the large, navigable rivers, as well as by the cultivation of the deep, rich soils.

The Farmington River in Simsbury and Connecticut River in Wethersfield have well-developed floodplains.

At any one time, a river may exhibit any combination of stream stages. A stream also may revert to a previous stage because of an uplift of the Earth's crust, change in climate, alteration in sea level, or other influences. Changes may take place swiftly, such as during a flood or earthquake, or over millions of years.

As discussed in Chapter One, Connecticut's bedrock valleys developed over hundreds of millions of years, and were modified and deepened by at least two glaciers. The modern stream system began to take shape during the meltback of the last glacier, between about 17,500 and 15,500 years ago. Stream courses and drainage patterns were influenced by the configuration of rock types and fracture zones.

The distribution of the glacial deposits that play a major role in watershed hydrology was also largely determined by the bedrock topography. Impermeable tills were left exposed on the hills, and more permeable stratified drift was deposited by, or in, meltwater, in the valleys.

Stratified drift consists of rock particles and fragments that were sorted by meltwater as a function of distance from the melting ice front. At each ice stand, the coarsest gravels and sands were deposited in contact with the ice. Medium-sized sand and gravels were deposited more distally as deltas, and fine sand and clays were deposited on lake bottoms farthest from the ice. Modern streams that flow across stratified

FIGURE 6

CHANNEL CROSS SECTION TERMS

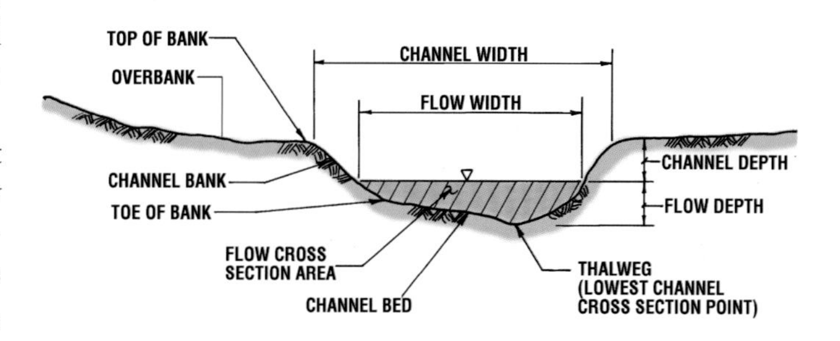

FIGURE 7

CROSS SECTION DEVELOPMENT

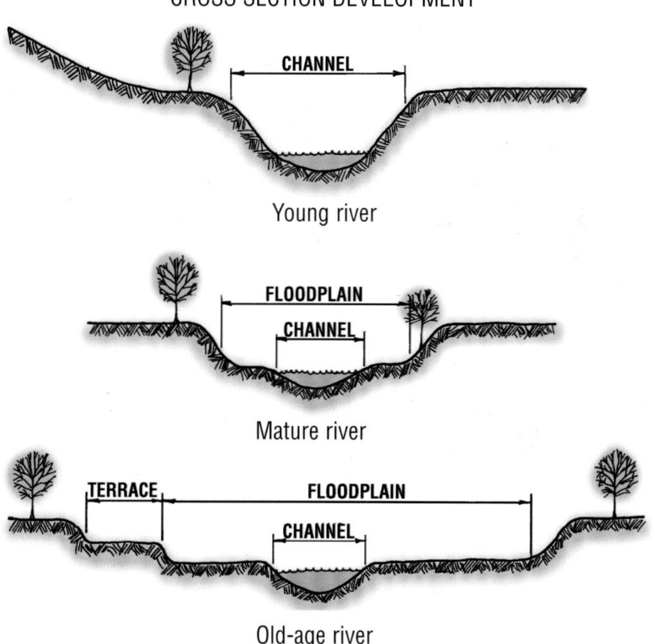

TABLE 12

SIGNS THAT A CHANNEL IS NON-ALLUVIAL

Bedrock is present	Alignment is fixed
Glacial till soils exist	No sedimentary floodplains exist
Artificial lining exists	Banks are fixed and stable
Little sediment activity occurs	

This channel is an example of a young mountain stream with steep slopes and exposed bedrock. Note the stair-like series of steep chutes and flat scour pools.

FIGURE 8

NON-ALLUVIAL CHANNELS

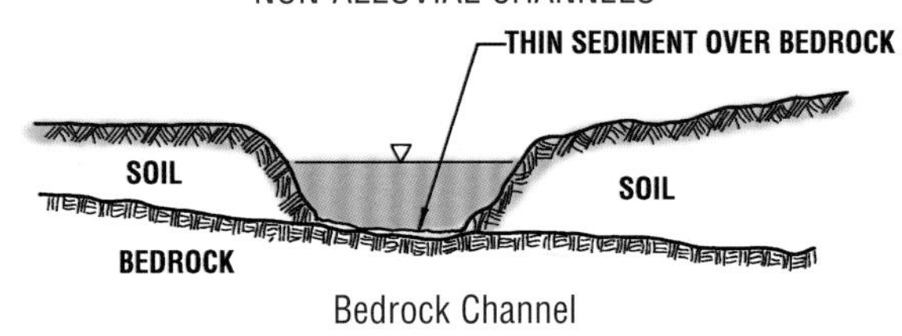

Bedrock Channel

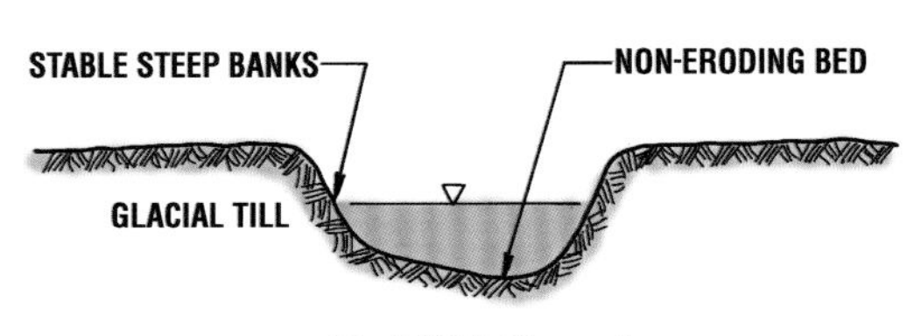

Glacial Till Channel

drift incised these deposits during the post-glacial upward rebound of Connecticut.

Stratified drift underlies many of Connecticut's valleys, where it is the main source of sediments transported by the state's major rivers. More significantly, it has a high capacity to store groundwater between its particles and is an important water-supply source.

NON-ALLUVIAL CHANNELS

It is important to recognize the difference between alluvial and non-alluvial rivers and their behavior (see Table 12).

Alluvial rivers flow through sedimentary deposits and usually have sandy beds. Since this material is easily eroded, the river's shape and size is determined by the river itself. This type of river is not confined.

Non-alluvial streams have beds and banks that are highly resistant to erosion. Underlying material may consist of dense glacial tills, old coarse sediment, or native soils formed by the decomposition of the bedrock.

The lack of channel erosion means these streams cannot easily adjust their channel size or location in response to their flows, and there is little sediment available for forming floodplains. Therefore, the shape and size of non-alluvial channels is dictated by the underlying channel material, not by the river's flow rate.

Within this category is the bedrock-controlled channel. As the name implies, it is so confined by bedrock that the latter determines the channel location and size. (See Figure 8.)

Man-made channels with non-erodible linings of rock riprap or concrete are also non-alluvial.

Mountain Streams

Youthful watercourses with steep slopes and rocky channels are often referred to as mountain streams.

Such streams can exist only on erosion-resistant materials, like boulders and cobbles, since their high flow velocities and turbulence would rapidly scour erodible soils. Mountain streams are usually rough, irregular, narrow, relatively deep, and lacking in floodplains.

The channel profiles are often stair-like, the water descending over ledge outcrops or boulders in short, steep chutes or falls. Beneath these, the force of the falling water digs out plunge pools. Sediment and smaller rocks swept out of the pools often accumulate at their downstream end.

Bedrock Channels

The size and shape of bedrock channels are controlled by the nature of the underlying rock.

Bedrock channels are prominent in eastern Connecticut and the western highlands, where the types of bedrock — predominantly schist, gneiss and granite — are hard and resistant to erosion, and also are found along the basalt ridges in the central part of the state. Characteristic features include waterfalls, rapids, gorges, and ravines.

In central Connecticut, most bedrock is a soft red sandstone that erodes rapidly. Consequently, few bedrock channels are exposed there. The limestone of western Connecticut also erodes easily, resulting in subdued valleys with little exposed bedrock, such as the Still River valley in Brookfield and New Milford.

Channels in Glacial Tills

The glacial till that covers much of New England has a strong influence on our river channels. Till was formed as glaciers scraped, mixed and smeared eroded material over the landscape. The resulting mix is made up of constituents ranging in size from microscopic clays and flour-like silt to large boulders. The material is usually dense, and because of its cohesiveness resists erosion.

Channels in glacial till tend to be relatively shallow compared with their width, with low, steep banks. The banks are usually stable and not subject to rapid erosion, especially where vegetation is established. Removal of vegetation increases the chance of soil loss.

Such channels usually do not have sandy "beaches" at the water line, since they lack suitable sediment.

Beds

A waterway that has carved a channel into glacial till or stratified drift deposits has eroded some of the bed material in the process. Small, lightweight particles are transported more easily by the streamflow than are heavy ones, resulting in a process that sorts the material by weight. As light particles are eroded and carried away, the heavier ones are left behind on the river bed. Eventually, the heavy particles accumulate, forming what is known as a lag deposit. (See Figure 9.)

The size of particles that remain depends on how fast the river flows and the composition of the original soil. For instance, areas with moderate velocities may have a bed of gravel, while higher velocities may result in a bed of cobbles and boulders.

Potholes scoured into a bedrock stream channel.

FIGURE 9

FORMATION OF AN ARMORED RIVER BED
BY SELECTIVE EROSION OF THE LIGHTER SOIL PARTICLES

1. Stream begins to erode channel.

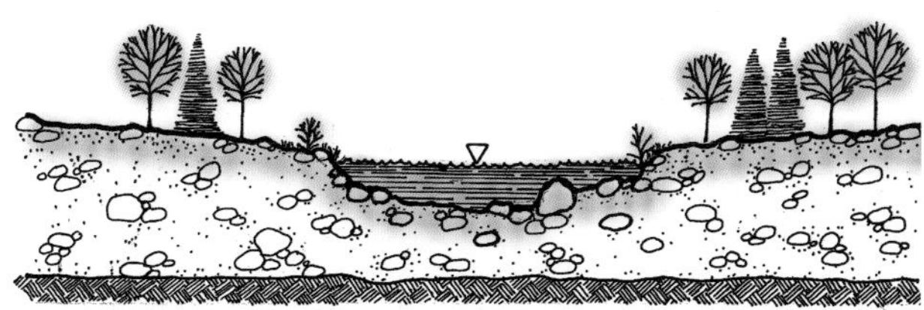

2. Lightweight soil particles erode, heavy rocks stay behind on channel bed.

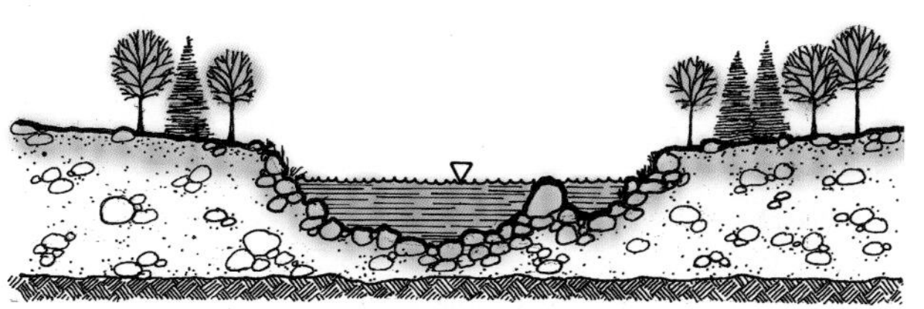

3. Vertical erosion stops when the accumulated heavy rocks cover and armor the channel bed.

An armored bed channel in a glacial till soil.

Static boulders armor this channel bed.

TABLE 13	
DYNAMIC ELEMENTS OF ALLUVIAL CHANNELS	
Shape and area of channel	Sediment load
Slope of channel bed	Streamflow rates and floods

A streambed covered with an extremely stable layer of material that is able to resist high flow velocities and floods is called a "static armored bed." The photo at left shows a typical lag deposit originating from glacial till.

Sometimes, the particles covering the bed, whether locally derived or sediments washed down from upstream, are stable under normal flow velocities but can be dislodged under flood conditions. The coarse surface layer of such a streambed, often composed of gravel, is referred to as mobile armor.

Because under normal flow rates the gravel bed limits vertical erosion, the channels of such rivers tend to be fairly wide compared with their depth. The water is often shallow, with large stones projecting above the surface. Portions of the Shepaug and Pomperaug rivers in western Connecticut and the Salmon River in eastern Connecticut are examples of this type of stream channel.

Under normal conditions, gravel bed rivers are non-alluvial. They become alluvial only when the bed material is subject to general movement, something that occurs only at high flow rates.

Because gravel-based channels can erode drastically if the thin armor layer is removed, special precaution is needed during construction projects to minimize the disturbance of armored streambeds.

ALLUVIAL CHANNELS

In contrast to non-alluvial channels, whose beds and banks resist erosion and change little over time, an alluvial channel is dynamic, its characteristics varying as the river erodes and redeposits the sediments of which the channel is composed (see Table 14).

Most alluvial rivers have beds of loose sand, which is easily eroded by water. But eventually an alluvial river establishes a state of quasi-equilibrium in which width, depth, and slope adjust to fit the normal flow conditions. When this occurs, the river is said to be stable. Even then, however, floods can mobilize its materials, transporting them downstream in suspension or along the bed.

Alluvial rivers are generally mature, and may have a single- or multiple-stem channel. Many are higher-order rivers with floodplains. Some smaller streams with low gradients also are alluvial.

Most of the alluvial or unconfined channels in Connecticut are incised into stratified drift deposits that choked or completely buried bedrock valleys during the melting of the last glacier. The volume of sediment produced by the glacier greatly exceeded the capacity of meltwater streams, and successive ice positions produced a series of glacial lakes, and associated deltas, in many of the prominent bedrock valleys. Some of these glacial sedimentary features were previously classified as Kane terraces. Prior to glacial unloading and upward rebound of the land, through-flowing drainage was poorly developed in these sediment-choked valleys.

In central and northwestern Connecticut, the more subdued limestone and sandstone valleys were particularly conducive to the formation of extensive glacial lakes, and predominantly fine-grained lake deposits completely buried the bedrock surface over wide areas. Here again, down-valley drainage was initially from lake to lake.

As Connecticut tilted upward to the north in response to glacial unloading of the land, stream gradients increased. Post-glacial streams began to downcut, and through-flowing drainage developed in the sediment-choked valleys and across the drained lake bottoms. Where this developing drainage cut previously deposited stratified drift, alluvial or unconfined channels developed. These channels appear underfit because they only cut through a portion of the glacially derived stratified drift.

TABLE 14

SIGNS THAT A CHANNEL IS ALLUVIAL

Floodplains created by sediment deposits
Sedimentary bed material
Sediment is actively transported by water
Banks and bed are subject to erosion
No bedrock is present

TABLE 15

NOTABLE EXAMPLES OF ALLUVIAL CHANNELS IN CONNECTICUT

- Upper Housatonic River, above Falls Village
- Quinnipiac River, Wallingford and North Haven
- Still River, Brookfield and New Milford
- Connecticut River, Wethersfield and Glastonbury
- Pootatuck River, Newtown
- Quinebaug River, Thompson to Griswold
- Farmington River, Avon and Simsbury
- Coginchaug River, Middlefield

CHANNEL BANKS

The streambanks are the sides of the depressed channel in which the normal streamflow is concentrated. The lower channel bank is the area below the ordinary water level and is too wet to support terrestrial vegetation. Aquatic plants, emergent wetland plants, and bare soil are common because this area is usually submerged and is subject to ice damage during winter. The upper channel bank is above the ordinary water level and generally can support upland vegetation and woody species.

Channels in unconsolidated sand and gravel material will tend to have mild slopes along the banks and "beaches" along the edge of the water. Eroding channels will tend to have steep banks. Material is removed from the bottom of the banks which will periodically slide into the channel when they are undermined and the slope becomes excessively steep.

In cohesive silt and clay soils, streambanks often have vertical slopes that suddenly collapse when they are undermined by scour or lose cohesion due to drying or frost. The collapse of vertical banks can deposit large blocks of soil in the channel that slowly break apart. The streambanks in cohesive materials along the outside edge of a bend

Deposits of coarse sand with ripple formations in an alluvial streambed.

TABLE 16	
CHANNEL SLOPES AND THEIR CHARACTERISTICS	
Very steep slopes	Channels with waterfalls, vertical drops, and cascades (more than 40 feet/mile)
Steep slopes	Channels dominated by fast runs and riffles with occasional pools, often with exposed boulders and cobbles (20-40 feet/mile)
Intermediate slopes	Channels with a mixture of half riffles and half pools, typically with a gravel and cobble bed (10-20 feet/mile)
Moderate slopes	Channels with visible flow velocities and a few mild riffles, usually corresponding to a silty or sandy bed (5-10 feet/mile)
Low slopes	Flat channels with limited visible water movement and no turbulence (0-5 feet/mile)
Impoundments	Lakes, ponds, and pools with flat water surfaces and little flow velocity
PROFILE ELEMENTS	
Falls	Water flowing over a vertical or near vertical drop, at least 3 feet high
Cascades	Steep channels with non-uniform, turbulent flow over boulders or bedrock
Runs	Long, moderately steep channels with higher velocities and some visible turbulence
Riffles	Shallow areas with turbulent flow between pools
Pools	Deep, wide areas with low mean velocities
Flats	Long, linear channel segments with flat slopes, low velocities, and no visible turbulence

will occasionally be undercut, with an eroded cavity extending beneath the bank due to concentrated flow. Undercut streambanks are usually a temporary feature except where reinforced by tree roots.

The channel banks along the ordinary water level are often irregular because of rocks, vegetation, stumps, and undercutting. The irregular edge helps to reduce flow velocities and provides shelter for aquatic life.

CHANNEL PROFILES

Profile refers to a river's downward slope, one of the most important factors both in a stream's hydraulics and in its ecology.

Rivers with steep slopes have high-velocity flow and are much more erosive than rivers with gentle slopes. The slope also affects a watercourse's alignment, width, depth, flow velocity, and its scour potential.

The channels of most lower-order streams do not have constant uniform slopes and may exhibit any combination of several distinct profile elements. (See Table 16.)

Cascades, Runs and Flats

Cascades occur where a stream channel is controlled by erosion-resistant material. They are usually found in steep channels lined with bedrock or boulders and are short in length. Cascades have turbulent whitewater flow but do not have vertical drops such as waterfalls.

Runs are long, moderately steep segments with high flow velocities. The bed has little loose material, often being composed of cobbles or bedrock. Especially steep runs may have boulders underneath. Their greater length and steepness differentiate them from riffles.

Flats are long, linear stream segments of uniform grade, where the flow is too slow to form riffles and the water not deep enough for pools. Flats often result when major floods wash out channel bars, or when bars are disturbed by humans. The lower photo on Page 33 shows a long, quiet flat along the Coginchaug River in Middlefield.

Pools and Riffles

The vertical profile of many natural channels includes a series of alternating shallow and deep sections in the streambed. During periods of low flow, the deeper sections are pools of nearly still water. Shallow areas are riffles of fast-moving water. This sequence may occur in channels that are straight or meandering. It is most pro-

A sequence of riffles, pool, and riffles.

FIGURE 10

POOL AND RIFFLE SEQUENCE

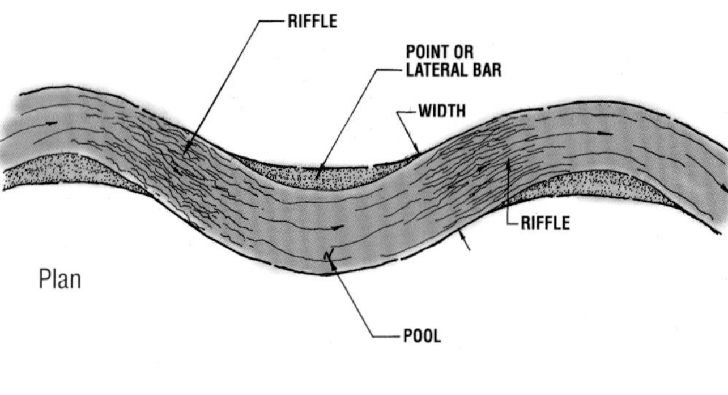

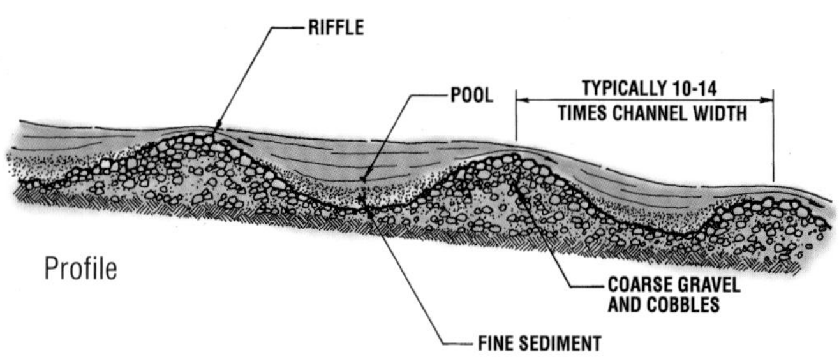

nounced where the riverbed consists of erodible coarse sand or gravel. (See Figure 10.)

Pools are deep and wide segments with currents slower than in adjacent areas. They are usually several times longer than the stream is wide. During times of normal and low flow, they generally have a bed of mixed grain sizes, including sediments finer than the mean.

During floods, pools are scoured by concentrated high velocity flows. Areas subject to scour are at the outside edge of bends and just downstream of sections where flood flow is concentrated by channel constriction, sand bars or other local conditions. During periods with low flow rates, the deeper pools have low flow velocities and provide shelter for fish during warm weather. In winter, they may have ice cover but do not freeze solid to the bottom.

Riffles are located downstream of pools, formed of the coarser sediment and debris scoured out of them. As the large, slow-moving volume of water in a pool encounters such an obstacle, it is forced over and down the far side in a shallow, more rapid flow. Flood flow velocities are high enough to remove fine grains, but not the coarser material, from riffle stretches.

The length of the pool is usually longer than that of the subse-

A flat with little bed slope and low flow velocity.

TABLE 17

CHANNEL PATTERN CHARACTERISTICS

Type	Attributes
Straight	Linear path
Curvilinear	Gentle, non-repeating, broad bends; slightly sinuous
Meandering	Repeating curves ranging from moderately sinuous to tortuous
Braided	Occasional bars or islands
	Overlapping bars or islands
	Multiple channels, many bars
Irregular	Random bends
	Limited bends
	Forced bends

quent riffle. The distance between the low point of a pool and the high point of the riffle usually is five to seven times the width of the channel. This ratio holds true for most rivers, large or small.

Riffles are important ecologically. Their turbulent whitewater mixes the water with air, raising dissolved oxygen to levels that allow fish to live. Several species require the oxygen-rich gravel beds as a place to incubate their eggs. Also, in northern climates, riffles are often the only surface waters that remain unfrozen during winter, providing drinking water for animals and feeding areas for birds of prey.

CHANNEL PATTERNS

Channel patterns and alignment are the result of both geologic history and current local conditions. The overall characteristics are the result of the geological processes that shaped the drainage basin and of major contemporary events such as floods. The current size and shape of a channel and floodplain are determined by past flow and sediment conditions. Smaller in-channel details, such as sand bars and scour pools, usually stem from recent conditions.

Bends

There are three types of river bends.

Forced bends occur where the river strikes erosion-resistant material. This may be bedrock, dense soils or the side of a valley. An extreme type of forced bend is a "fixed bend," which makes a river change course at a sharp angle. Only slowly, over a long period of time, is a fixed bend eroded into a curve.

Limited bends occur where river banks are somewhat resistant to erosion. Underlying materials may be glacial tills or sedimentary deposits too coarse to be easily moved by present flow. Limited bends are often characteristic of entrenched rivers that erode their beds vertically with little lateral movement.

Free bends occur where banks are composed of unconsolidated materials that erode easily. They are frequently found along alluvial channels on floodplains.The geometry of limited and forced bends is irregular, depending on the degree to which erosion has smoothed a curve. The stability of free bends and their rate of lateral bank erosion appear to be related to the rate at which material is eroded from the bottom and the radius of the bend. (See Figure 11.)

Types of Patterns

The five basic channel patterns are irregular, straight, curvilinear, meandering, and multi-channel braided. (See figure 13.)

Different segments of the same river may have different channel patterns. The pattern of a given segment may also change from one year to the next, depending on flow rates, sediment loads, sediment size, and upstream and downstream conditions. Occasionally, patterns may overlap; for example, a braided stream may also meander.

The local pattern is influenced by the slope, sediment, and water flow. For example, the converging flow that occurs at river bends tends to scour the bed and banks, while the diverging flow that occurs where the channel width increases tends to leave sediment deposits. (See Figure 12.)

Irregular Channels

As the name implies, irregular channels have a variable alignment without a dominant pattern. They are almost always non-alluvial channels whose alignment is controlled by non-erodible glacial till or bedrock. They have numerous forced bends with a variety of irregular angles. They include the mountain streams found in areas with rugged topography, as well as watercourses incised within erosion-resistant banks or young streams that are still cutting downward through easily erodible soils.

Straight Channels

Channels are considered straight when their sinuosity is low, with only minor bends along their course. The few bends may have a repeating or random pattern. During low flow, one can often see pools and riffles.

Straightness occurs in response to two conditions. The first is when the prevailing slopes are flat and the flow is so slow that it does not erode the banks. The second is when slopes are steep and the flow is swift enough to erode a straight path, such as in a gully or ravine.

The stability of a straight channel depends on how well the soil resists erosion where the thalweg, the deepest part of the streambed, traverses the outside banks of the pools. If the flow is fast enough to erode the bank, the channel will evolve into a meandering or irregular alignment.

The length of straight channels in erodible soils is seldom longer

FIGURE 11

CHANNEL MEANDER PATTERNS: EXAMPLES OF FREE BENDS

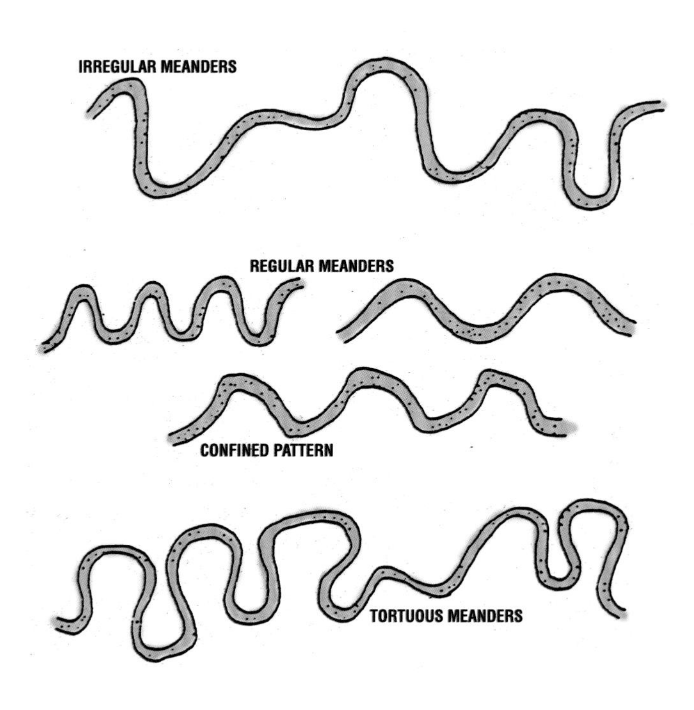

Meanders may occur in streams of any size.

FIGURE 12

RIVER FLOW TYPES

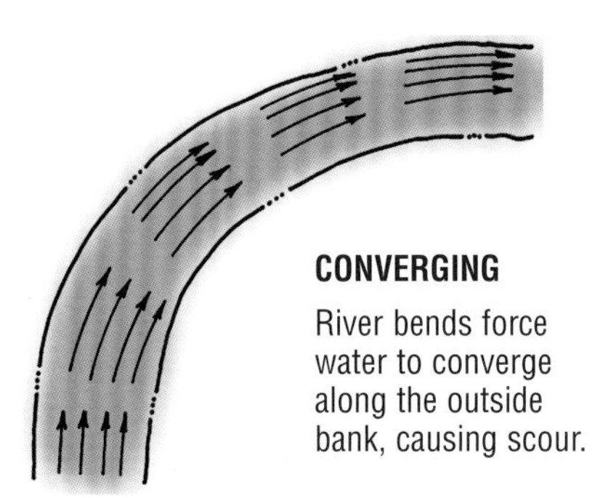

CONVERGING

River bends force water to converge along the outside bank, causing scour.

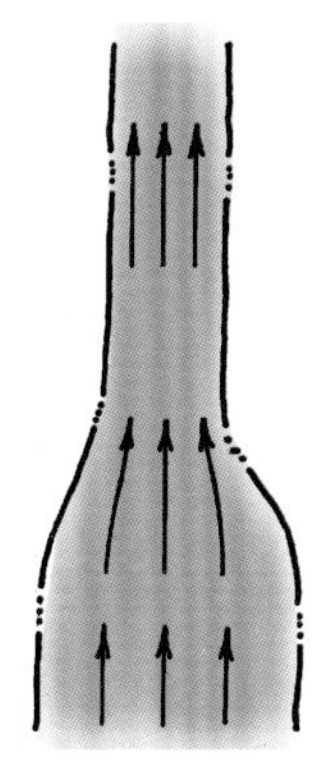

CONTRACTION

Decreasing channel width concentrates (contracts) the flow, causing scour and higher velocities.

EXPANSION

Increasing the channel width decreases velocity as the flow expands outward, encouraging sediment deposition.

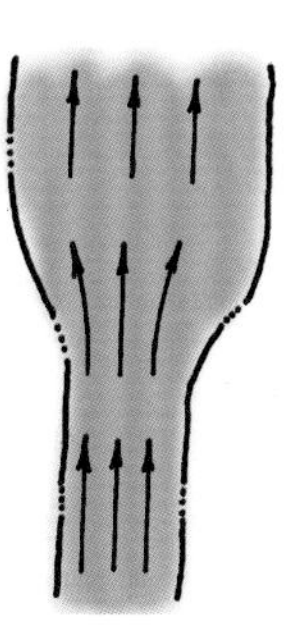

FIGURE 13

RIVER CHANNEL PATTERNS

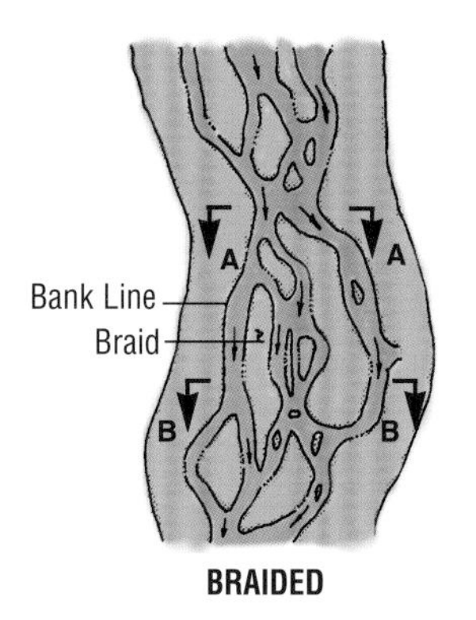

BRAIDED

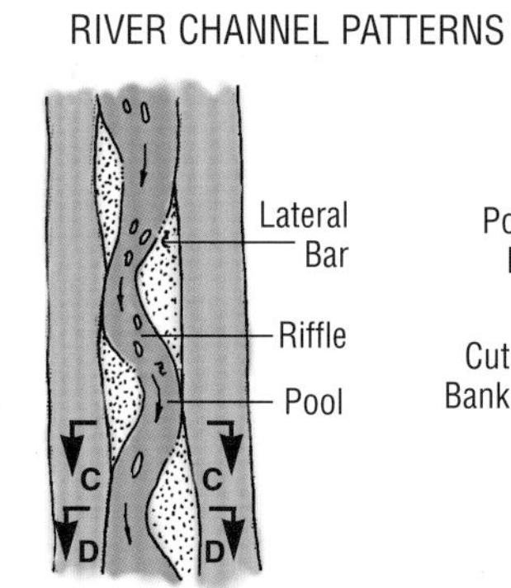

STRAIGHT

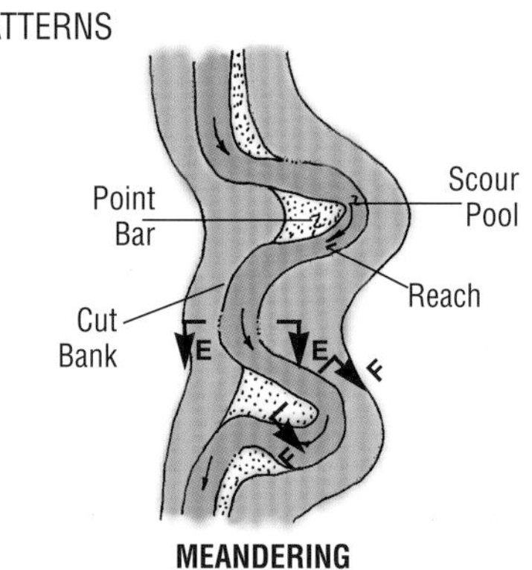

MEANDERING

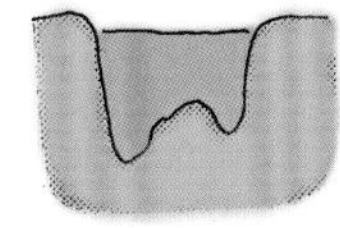

SECTION A-A

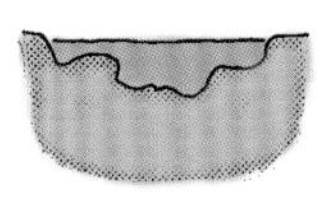

SECTION C-C

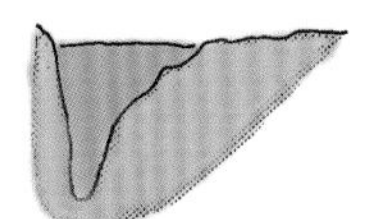

SECTION E-E

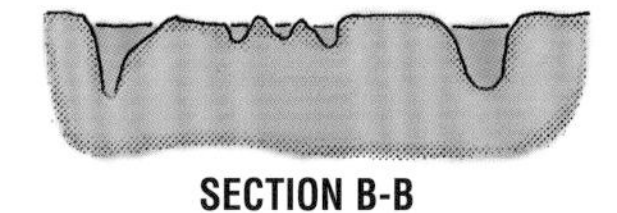

SECTION B-B

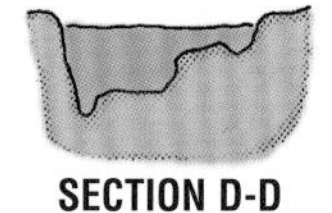

SECTION D-D

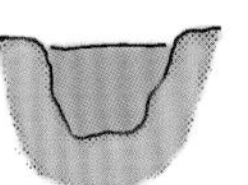

SECTION F-F

Modified from U.S. Geological Survey

A small, meandering stream. Note the point bar of gravel at the inside of the bend and the undercut, eroded bank along the outside of the bend.

than 10 times the width of the flood flow, because non-uniform flow, irregular bank vegetation, and changes in bed material will usually induce bends or even meanders.

Straight channels are often only temporary, such as the "runs" that form between meanders. Others reflect human activity, such as dredging or measures to stabilize banks.

Curvilinear Channels

Curvilinear channels are between straight and meandering in appearance. They have long, gentle curves that lack the regular, repeating pattern of meandering rivers. They generally have mild to moderately steep slopes with narrow or erosion-resistant floodplains that restrict the development of broad meanders. Pools and riffles are common.

Curvilinear channels are frequently misfit streams that are entrenched in an earlier floodplain. (See section on Misfit Streams.)

Meandering Rivers

One of the most common patterns for alluvial rivers is a meandering alignment. This term is reserved for rivers with a distinctive, repeating, sinuous pattern of curves.

The classical meandering river has evenly spaced, symmetrical curves along its length. (See Figure 14). It is usually narrow and deep, with pools forming at the bends and riffles occurring midway between them.

One of the most notable characteristics of a meandering river is the channel's ability to shift laterally across the floodplain. This happens as the flow is concentrated on the outside of the bends and the banks there are eroded. Sediment tends to be deposited on the inside of the bends, where the flow velocity is slower, forming a point bar. As the channel shifts outward at each bend, the bar expands to fill the old channel. (See Figure 15.)

The thalweg, or lowest channel point, of meandering rivers is usually found along the outside of the bends, where most of the scour occurs. Thus the thalweg shifts from one bank to the other following the meander pattern.

Meanders may also move downstream. This occurs when the flow maintains its momentum through the lateral part of the bend and erodes the concave bank just downstream. Point bars also tend to

FIGURE 14

CLASSIC MEANDER DIMENSIONS

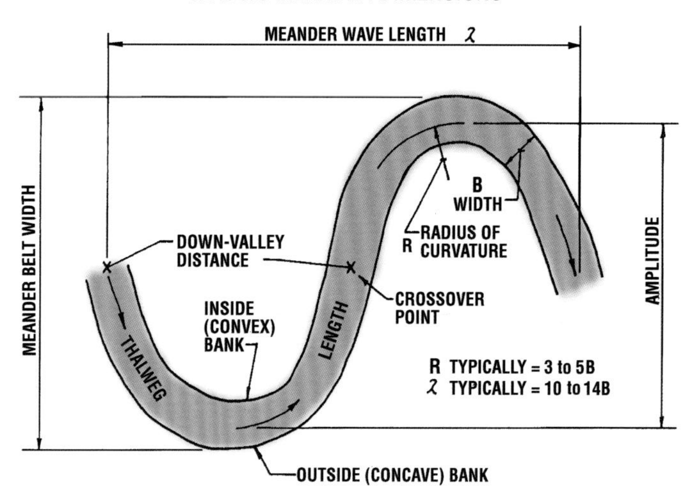

FIGURE 15

MEANDER BEND DYNAMICS

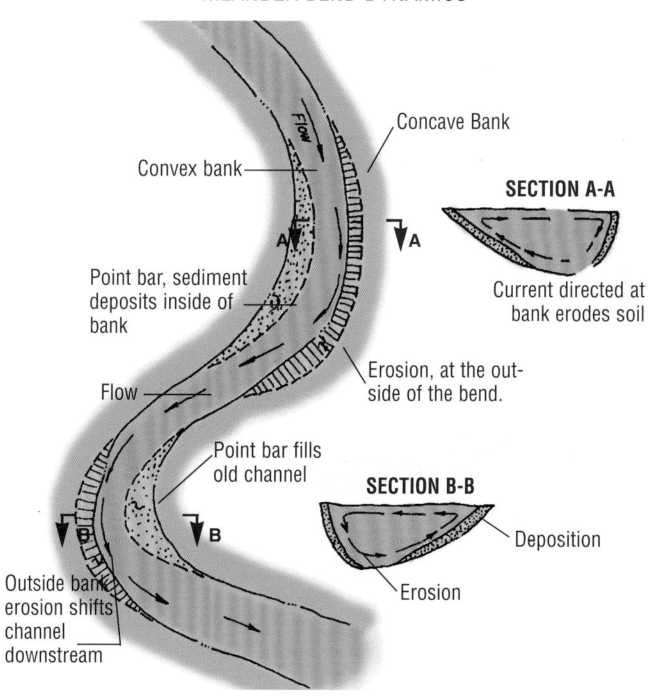

TABLE 18	
SINUOSITY	CHANNEL/VALLEY LENGTH RATIO
Low	1.0 - 1.3
Moderate	1.3 - 2.0
High	Greater than 2.0

An alluvial river with sinuous meanders.

form in the downstream direction as the material eroded at the bend is deposited in the relatively still water on the downstream side of the existing bar.

The fact that alluvial river channels can move is one of the key arguments against development of floodplains. Since a river can change its course, the traditional practice of constructing buildings above the historic high water mark is not always sufficient.

In mature, stable rivers, the classical meander length is a function of three variables: stream width, discharge, and sediment size. The meander length is usually from 11 to 16 times the channel width. The curve radius in mature alluvial rivers is typically two-and-one-half to five times the channel width. (See Figures 14 and 15.)

A river's sinuosity is gauged by the ratio of the channel length to the length of its valley. The length of a channel, when measured along its center line following all of the curves, is always longer than the valley in which it lies. High sinuosity is usually associated with mature rivers with very mild gradients (see Table 18).

Examples of meandering channels include the Quinnipiac River in Wallingford and North Haven, the Connecticut River between Rocky Hill and Hartford, the Housatonic River north of Falls Village, and the Still River in New Milford.

Braided Rivers

Braided rivers are wide and shallow with a series of smaller, interlaced channels separated by numerous bars and islands. Such rivers may be straight or may curve, and generally form where water flowing down steep slopes over erodible soil carries a heavy load of coarse-grained sediment.

The dominant features of braided rivers are the bars and islands of sediment deposited by high flows in the center of the channel.

The bars grow in a downstream direction as more material is added in the wake of initial deposits. They also widen slowly, deflecting flow outward toward the banks and causing the river to erode laterally. Bars that extend above the water surface and can support vegetation eventually become islands.

Braided channels are unstable because high flows can shift the position and size of bars. Pools and riffles can form temporarily or more permanently. Human activities that create sediment or increase deposition can also affect these streams.

In Connecticut, natural braided rivers are rare because of limited sediment sources. Generally, the underlying rock here — schist, gneiss, granite, and basalt — is too hard to erode readily, and most of the rock that is soft by nature (sandstone, limestone) was long ago eroded below current floodplain levels.

Local examples of braided channels include portions of the Naugatuck River where dams have altered natural flow and sediment loads, and portions of Meetinghouse Brook in Wallingford altered by the 1982 flood.

Other sites include:

- Small streams with high sediment loads from nearby farms or construction sites.
- Streams flowing across alluvium in old, man-made lake beds behind breached dams.
- Gravel extraction sites.
- Coastal estuaries.

Effect of Slope on Pattern

For a given discharge, alluvial channel patterns vary with slope and sediment load.

On flat terrain, channels tend to be narrow and deep, with low flow velocities and straight thalwegs. On steep slopes, channels tend to be straight because high-velocity flood flow tends to rush straight down the valley, bypassing what bends may exist and eroding a new channel across them.

Figure 16 shows the influence of slope. Note how straight channels usually have flat or extremely steep slopes, while meandering channels have moderate slopes.

- The pattern of an alluvial channel at equilibrium is influenced by its discharge rate and bed slope.
- An increase in slope can change the pattern from straight to meander, or from meander to braided.
- For a given slope, a braided channel has a higher flow capacity than a meander channel.

CHANGES IN ALIGNMENT

Rivers, especially alluvial ones, may change the alignment and location of their channels as a result of floods, erosion, sediment deposits, or human activities.

TABLE 19

BRAIDED RIVER CHARACTERISTICS

- Multiple flow paths and channels
- Excess sediment loads
- Moderate to steep slopes
- Sediment bars
- Islands

FIGURE 16

CHANNEL SLOPES AND PATTERNS

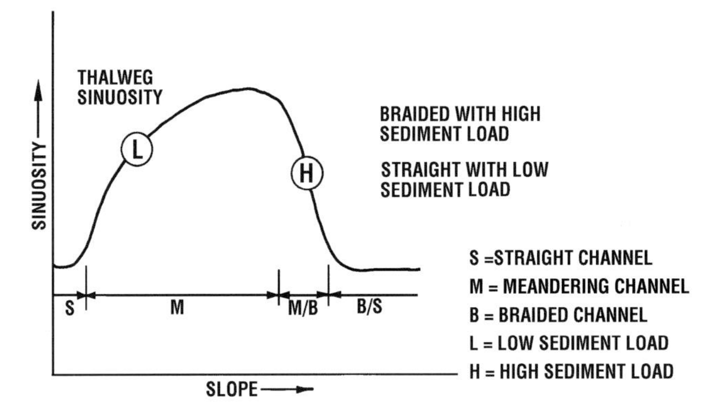

Rocky beds and steep slopes are typical of mountain streams.

FIGURE 17

CONNECTICUT RIVER ALIGNMENT CHANGES — GLASTONBURY MEADOWS

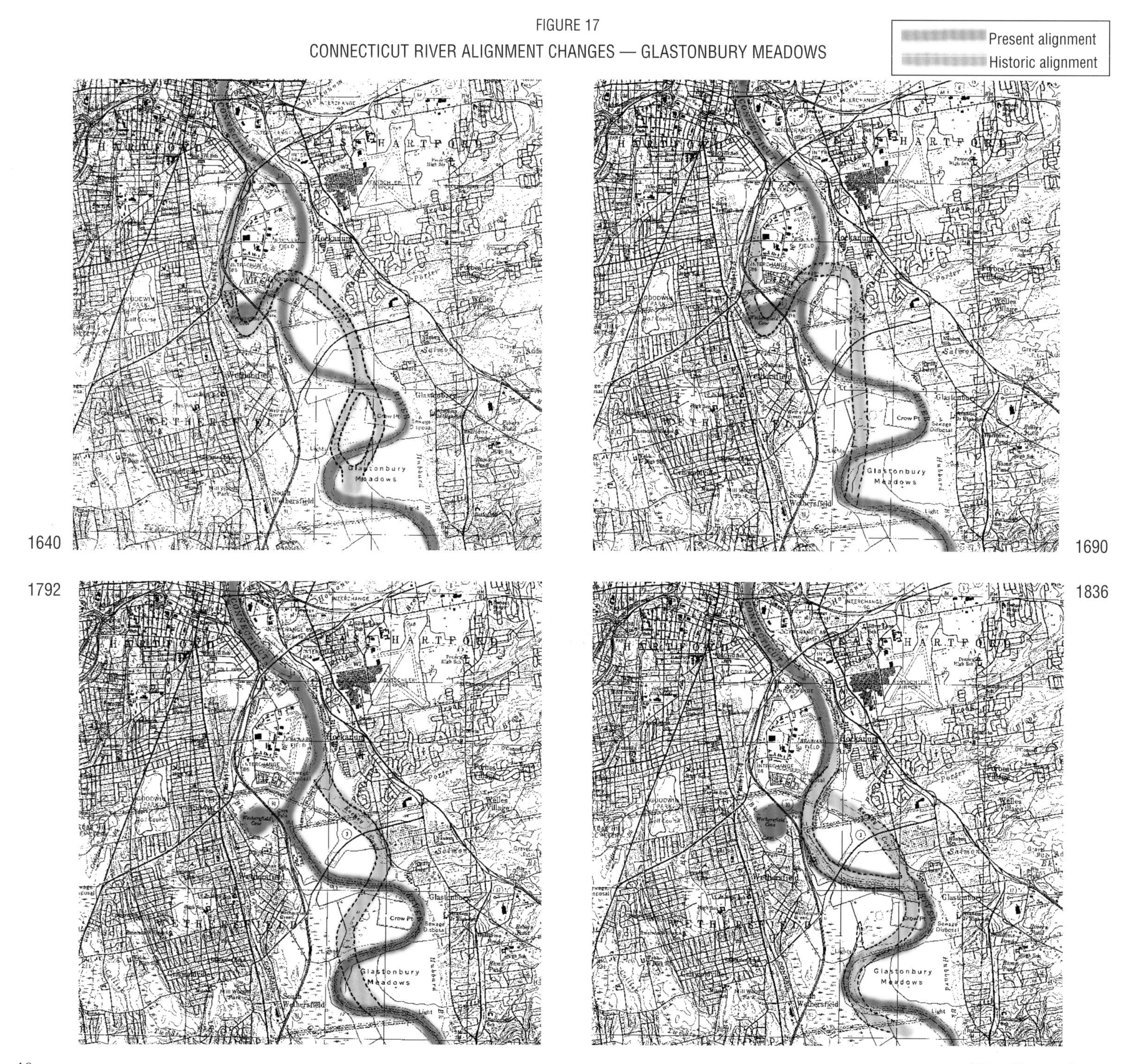

The rivers in Connecticut are considered stable compared with those in other geologic areas. But there is evidence that some of our channels are changing.

The Still River, for instance, has a tortuous meandering alignment as it flows toward the Housatonic River. In one quarter-mile-long section alone, 13 major changes in alignment occur. Numerous oxbow lakes, formed from sections of the channel that the river has since bypassed, testify to the changes in alignment. (See Figure 18.) Similarly, the Farmington River in Avon and Simsbury has an active meandering alignment (as can be seen on the U.S. Geological Survey's Tariffville Quadrangle Map).

A portion of Roaring Brook in Glastonbury altered course after a flood in 1979. Also, sections of Meetinghouse Brook in Wallingford changed from a straight to braided channel after the June 1982 flood.

During the past 300 years, the Connecticut River has had major alignment changes between Hartford and Rocky Hill. (See Figure 17.) Early maps show the river originally entered Wethersfield Cove near the present day Folly Brook. It then swung northeasterly, before flowing south through what is now called Keeney Cove in Glastonbury. The area now known as Crows Point was originally Wrights Island, in the mid-channel of the river.

The Connecticut's course changed radically about 1700 as the result of a flood. The river bypassed the old Colonial port on Wethersfield Cove and abandoned Keeney Cove, meandering across the meadows. As the channel changed, Wrights Island became part of the Wethersfield shore.

Today the banks are stabilized with stone riprap placed by the United States Army Corps of Engineers. However, humans' attempts to restrain the river may not be sufficient to prevent future changes.

FLOW VELOCITY

The average flow velocity in rivers varies with slope, flow rate, channel shape, and friction roughness. The bed material in the channel is closely related to flow velocities (see Table 20).

The velocity of water tumbling down steep headwater and mountain streams is often deceiving. Although these streams may have steep slopes, the extensive contact between the shallow flow and the rocky beds and protruding boulders creates high friction levels that slow the water.

FIGURE 18

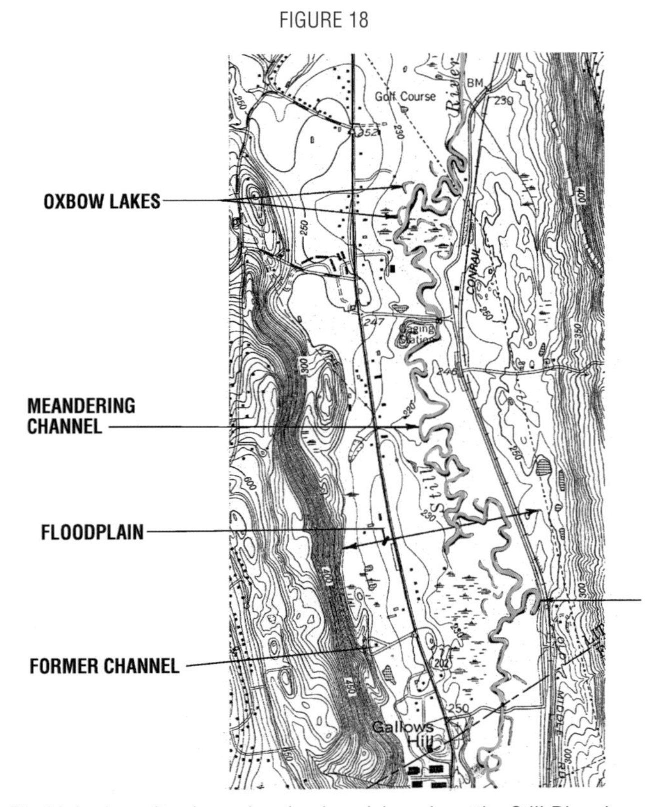

The high-sinuosity channel and oxbow lakes along the Still River in New Milford result from the very flat slopes of the easily eroded limestone valley.

An alluvial channel (gravel, mobile-armored) with a braided pattern.

TABLE 20

HOW FLOW VELOCITY AFFECTS RIVER BEDS *

Velocity (Feet per Second)	Dominant Channel Bed	Habitat
>4	rocky	torrential
>3	heavy cobble	torrential
>2	light cobble	non-silted
>1	gravel	partly silted
>0.7	sand	partly silted
>0.5	silt	silted
<0.5	mud	pond-like

* Modified From: Tansley, A.G., "The British Islands and Their Vegetation," Cambridge University Press, England; 1939.

Braided channels, such as this short reach, occur in response to the accumulation of sediment.

TABLE 21

SEDIMENTS AND THEIR SIZES

Clay	Microscopic
Silt	Similar to ground flour
Sand	Similar to fine to coarse sugar
Gravel	Size of corn kernels to golf balls
Cobbles	Size of baseballs to volleyballs

The velocity during normal streamflow conditions usually ranges between one and three feet per second, a typical walking speed. At this velocity, there is little erosion or sediment deposition for sandy soils.

Flow velocities in larger rivers can be surprisingly high. Although large channels usually have mild slopes as they approach the lower end of their watersheds, they still can have appreciable flow velocities as their deep waters flow with little friction over smooth, sandy beds.

Few rivers have sustained high flow velocities, because such velocities are very erosive and would eventually reduce the channel slope. Flow velocities of more than 10 feet per second are rare. Common flood flow velocities range from five to 10 feet per second in channels, with lower speeds on floodplains.

Flow velocities are not uniform within a channel. The highest velocities in straight channels are usually near the center, where there is little friction, and often are up to 50 percent higher than mean velocities. In curved channels, higher flow velocities occur along the outside edges of the bends.

THE SEDIMENT CYCLE

The sediment cycle starts as soils in the watershed erode and are transported by surface runoff that washes into rivers. Subsequent movement of sediments through the river system to the sea is a complex process. Generally, it is made up of many cycles of scour, movement, transport, and deposition. Heavy particles, such as gravel and cobbles, usually originate in the channel itself. Lighter, suspended particles of silt, clay or sand may originate in uplands or be products of channel erosion (see Table 21).

Sediment movement occurs when water flow exerts sufficient force to overcome the resistance produced by the weight of individual particles, their cohesion to similar particles, and their friction with the streambed.

When the force of the water is the maximum possible without causing sediment movement, it is said to be at its permissible velocity. This is difficult to measure, since not all particles begin to move at the same time.

Most sediment is transported during periods of high water flows and high velocities. High flow velocities are able to erode and transport larger particles than low velocities, and so accelerate erosion. Similarly, long-duration floods can cause more erosion and sediment transport than short-duration floods.

During the March 1936 flood, sediment deposits on the Connecticut River floodplain ranged from 1.25 inches in thickness to 6 feet, according to United States Geological Survey records.

Broad, flat floodplains are excellent agricultural land because of the lack of rocks and the periodic deposition of nutrient-rich sediments.

Sediment Loads

Suspended sediment consists of small particles that are transported as part of the water mass. Bed-load sediment consists of heavier particles that are pushed, dragged, and bounced downstream along the bed of the channel.

Portions of the sediment load are deposited in the channel or on the floodplain, in several forms. They may be temporary or semi-permanent. The various types are noted in Figure 19.

The sediment concentrations in river water and long-term sediment loads depend on the availability of erodible soil and the ability of a river to transport it. There are many types of sediment deposits, some of which contribute to the formation of floodplains (see Figure 20).

FIGURE 19
TYPES OF SEDIMENT LOAD

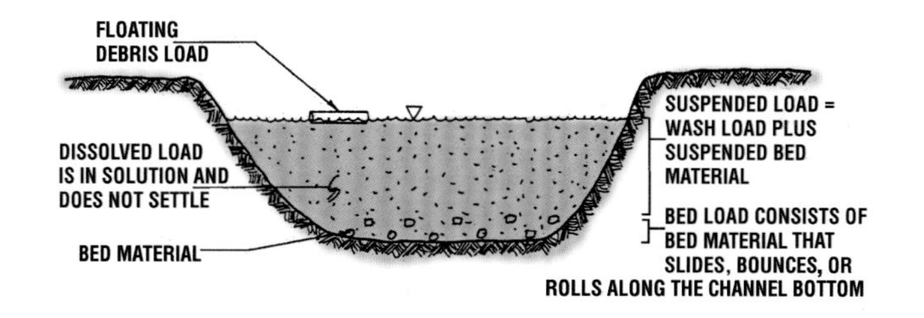

Sediments in Connecticut Rivers

Under natural conditions, Connecticut's streams have low sediment loads and concentrations of suspended sediments compared with rivers in other regions, notably the Western states. This is a result of the state's erosion-resistant soils and the humid climate with dense vegetation, which retards erosion.

Although information on suspended sediment loads in Connecticut is limited (there is little base-line data on bed loads), some insight is provided by United States Geological Survey reports on suspended sediment loads in two eastern Connecticut streams, the Yantic River and Muddy Brook in Woodstock.

The Yantic River data, collected from October 1975 to September 1980, showed a total five-year sediment load of 19,000 tons.

This is equivalent to an average daily load of 13.4 tons for the rural, 98-square-mile watershed, and indicates that the basin's average annual sediment yield is between 50 and 60 tons per square mile of watershed. These findings are similar to data obtained from 19 watersheds in Massachusetts.

The data show extreme variations in sediment load depending on flow conditions.

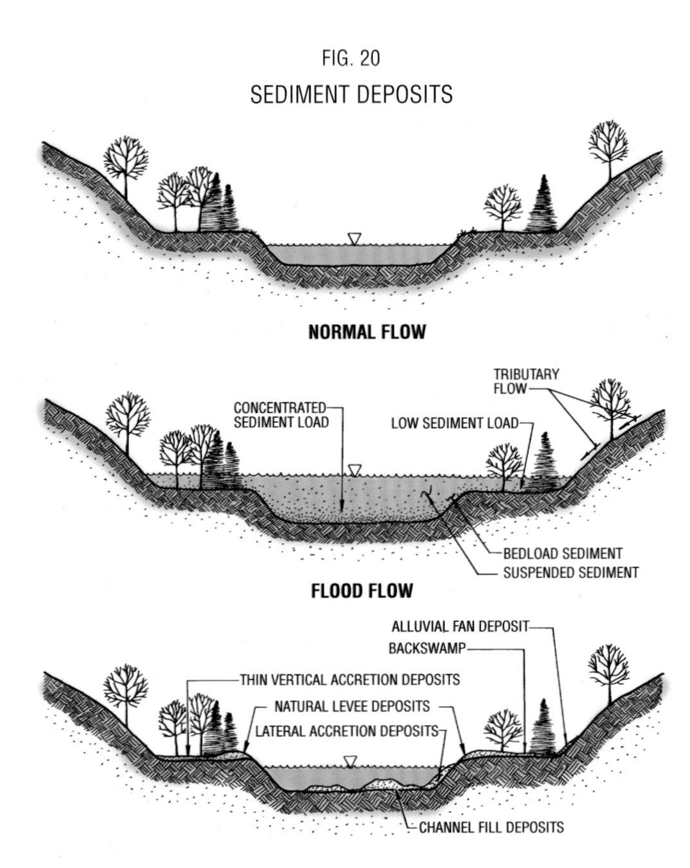

Post-Flood Deposition Pattern

During periods of average flow, the concentration of suspended sediment was usually three milligrams per liter or less, and the river's sediment discharge less than one ton per day.

Concentrations and discharges both increased sharply during periods of heavy rain, and just four storms were responsible for 48 percent of the total sediment discharge during the five-year period. The concentration of suspended sediment exceeded 10 milligrams per liter only 10 percent of the time, varying up to 168 milligrams per liter. (A concentration of 50 milligrams per liter gives water the color of chocolate milk.) The greatest sediment loads were associated with spring storms, when agricultural fields were fallow without ground cover.

During a typical rainfall, the sediment concentrations increase rapidly as streamflow increases. They peak and begin to recede only when the streamflow tapers off. Concentrations decrease rapidly at first. They then begin to recede at a slower rate, until pre-rain concentrations are again established.

The Muddy Brook data were obtained from 1980 to 1983. The results are similar to those for the Yantic River, with a mean suspended sediment concentration of 6.4 milligrams per liter and a range of zero to 163 milligrams per liter. The measured sediment yield was 32.5 tons per square mile per year.

FLOODPLAINS

Geologists define a floodplain as the area covered with sediment deposited by floods. The hydrologist, in contrast, defines it as the area that is periodically inundated by water during floods. The National Flood Insurance Program identifies floodplains as those areas inundated by a 100-year flood, disregarding geologic factors. Although the definitions vary, they all take into account the land's being subject to flooding.

Alluvial floodplains, using the geological definition, are one of the most important landscape features.

Their level topography and proximity to rivers makes them excellent locations for development.

Being composed of rich sediment and free of stones, floodplains are highly valued for agriculture. Even where floodplains are subject to annual spring floods, they are excellent for growing late-season crops.

Floodplains also have crucial natural functions. They convey excess runoff and temporarily store water in times of flooding, thus reducing peak flood flows and velocities in the river channel. They can also hold groundwater, releasing it back to the channel and supplementing streamflow during extended dry periods.

The Farmington River is an interesting example of how a floodplain can store and delay surface runoff, thereby reducing peak flows. Records from United States Geological Survey stations show that during the flood of August 1955, the peak flow actually declined as the water passed downstream from Collinsville to Rainbow Dam in Simsbury, despite the increase in the size of the watershed area. This reduction resulted because waters were spread across the floodplain in Farmington, Avon, and Simsbury. Unfortunately, much of that area was developed and suffered flood damage. (See Table 22.)

TABLE 22

FARMINGTON RIVER AUGUST 1955 PEAK FLOWS

GAUGE LOCATIONS	WATERSHED AREA (SQUARE MILES)	PEAK FLOW RATE (CUBIC FEET PER SECOND)
New Boston, Mass.	131	57,200
Riverton	217	101,000
Collinsville	360	140,000
Rainbow Dam	589	69,200

Floodplain Formation

Rivers create their flat floodplains by eroding and depositing excess sediment, primarily during floods. The wandering meander patterns lead to lateral erosion that reshapes the sediment deposits, also called alluvium. Since channel size tends to remain constant for a given discharge, erosion of one bank leads to the deposit of sediment on the opposite one. In this manner, a river can shift its position from one side of the valley to the other. This process widens the valley, eventually creating a plain.

The floodplain is higher than the river itself during periods of normal flow, but lower than the elevation of floodwaters.

Floodplains are usually associated with mature, large, meandering rivers. But even a wandering, straight river can create a floodplain. (See Figure 21.)

FIGURE. 21

ALLUVIAL FLOODPLAIN FORMATIONS

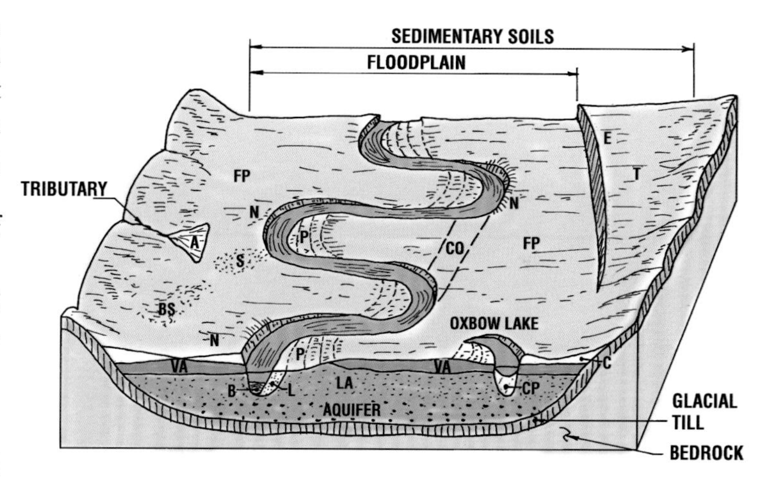

TYPICAL ASSOCIATIONS OF VALLEY SEDIMENT DEPOSITS

VA	Vertical accretion	**A**	Alluvial fan
N	Natural levee	**B**	Bedload deposits
FP	Floodplain	**L**	Log deposit
BS	Backswamp	**F**	Channel Fill
LA	Lateral accretion	**C**	Colluvial
P	Point bar deposits	**CO**	Meander cut off (chute)
S	Splay	**T**	Floodplain terrace
E	Escarpment	**CP**	Clay Plug

Modified from American Society of Civil Engineers Manual #54 "Sedimentation Engineering," 1975

Floodplain Features

Channel fill deposits occur when fill rates exceed scour rates and the river is unable to convey the full load of sediment washed into it from its headwaters and tributaries. It is forced to deposit some of the material in the channel bed, gradually raising it and sometimes forcing the water to shift toward the sides. Mid-channel deposits often form sandbars or islands. The material is poorly sorted, with a wide range of particle sizes that are placed irregularly and rounded from abrasion.

Lateral accretion deposits are formed along the edge of the channel, particularly in the form of point bar deposits on the inner side of

FIGURE 22
OXBOW LAKE FORMATION

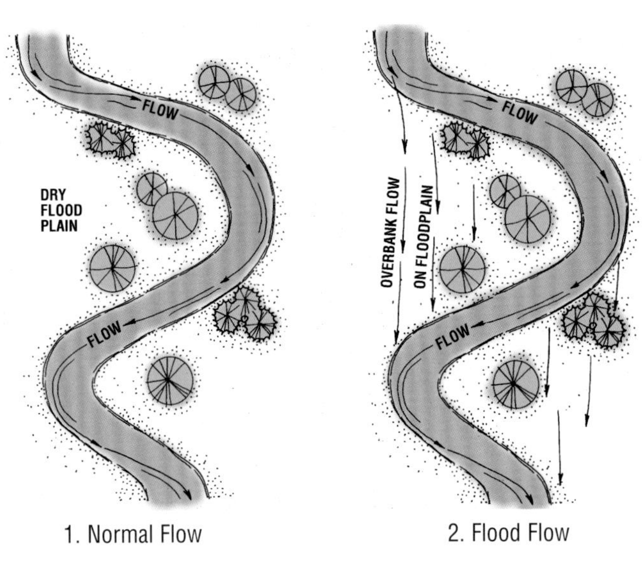

1. Normal Flow

2. Flood Flow

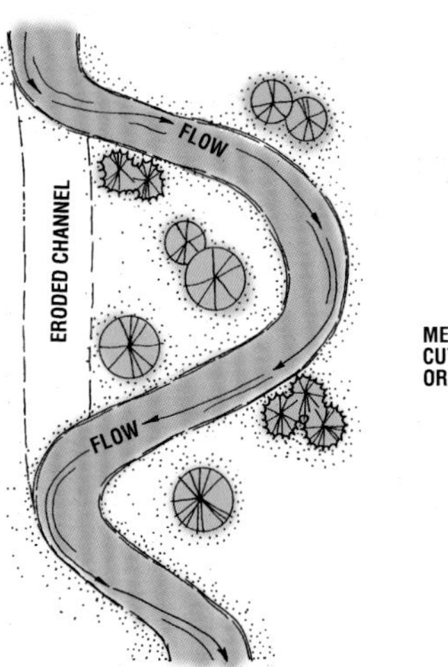

3. Post-Flood

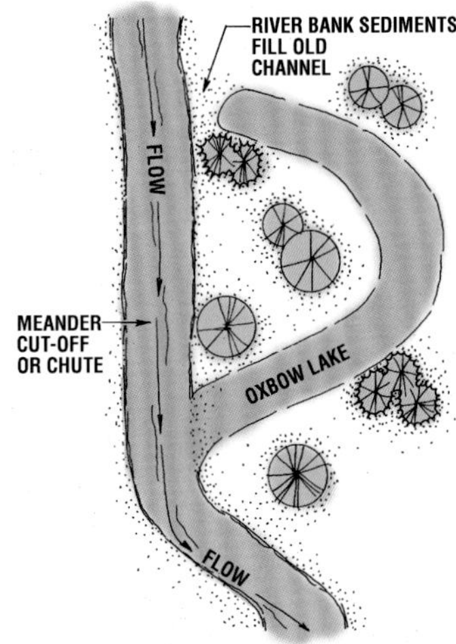

4. Oxbow Lake

meander bends. They are made of coarse sediments, frequently covered with finer material from subsequent overbank flows.

Channel lag deposits are composed of very coarse sediment material that accumulates in the streambed. They are usually found in steeper reaches or riffles, where the finer, lighter particles have been washed away by selective erosion. The lag deposits often "armor" the bed of the channel and slow the subsequent rate of bed erosion. Coarse particles are left behind when a channel shifts its course. They are sometimes covered with lateral accretion deposits generated by the shifting channel.

Vertical accretion deposits are the results of overbank flood flows that are so slow they do not convey all of the sediments downstream. The finer particles are carried away from the channel and deposited in stratified layers. The deposit is almost entirely from the suspended sediment load and may be high in fine sand and silt.

Floodplain splays are fan-shaped deposits of coarse material left at points where the river flow first overtops low points along the banks. They are found at and behind breaks in the natural levees along the top edge of the channel.

Colluvial deposits occur along the edge of the floodplain at the bottom of the valley walls. They consist of material deposited by upland sheet erosion and talus from degrading valley sides. Often they include angular pieces of bedrock and till that has not been sorted by flowing water.

Natural levees are deposits of coarse sediments that settle and accumulate on river banks as water overtops and leaves the channel, losing velocity as it enters vegetation. Such material forms natural levees higher than the rest of the floodplain.

Backwater swamps are wetlands created by surface water trapped on floodplains by natural levees that prevent water from returning to the river.

Meander chutes are secondary flow channels that convey overbank floodwater between meander bends. If a meander chute erodes deeply enough, it will convey water during normal flow, bypassing the meander and becoming a new channel. In the process, it creates an oxbow lake out of the abandoned meander.

Alluvial fans are sediments deposited by tributaries on the top of the floodplain.

Among other, less common deposits are valley plugs. These form in abandoned channels that — after being blocked by debris, drift-

FIGURE 23

ALLUVIAL RIVER FLOODPLAIN FEATURES

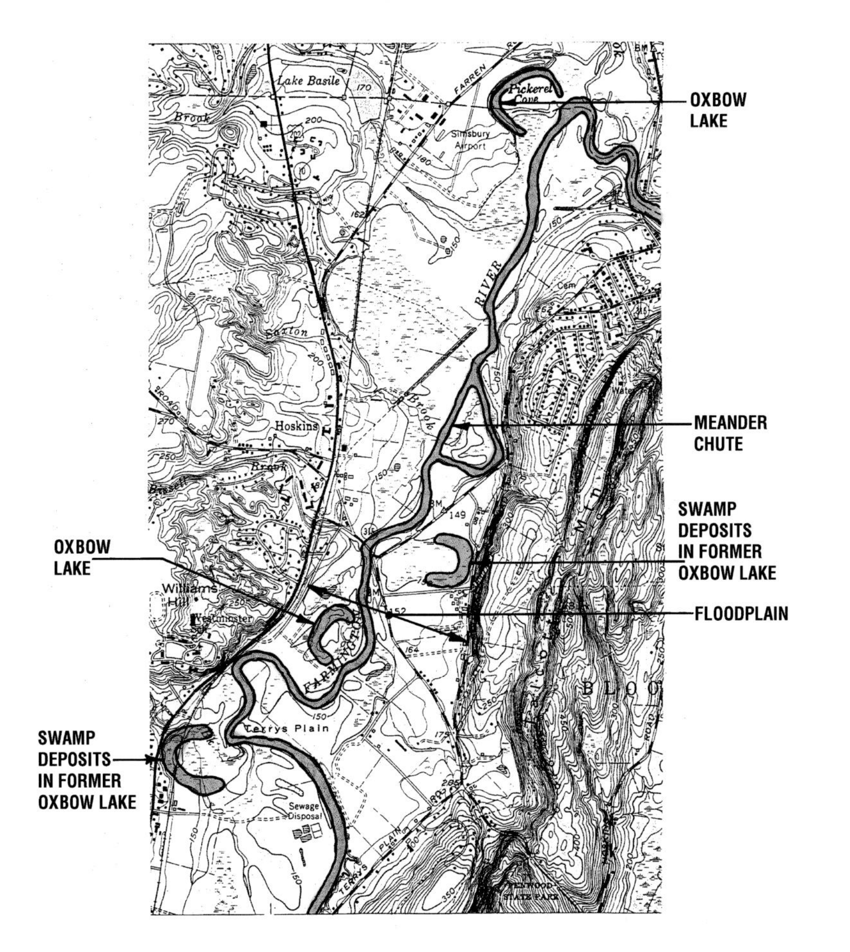

At right, an oxbow lake that formed in an abandoned meander of the river at left.

The deposits of fresh white sand along this river bank are forming a natural levee.

FIGURE 24

FLOODPLAIN & TERRACES

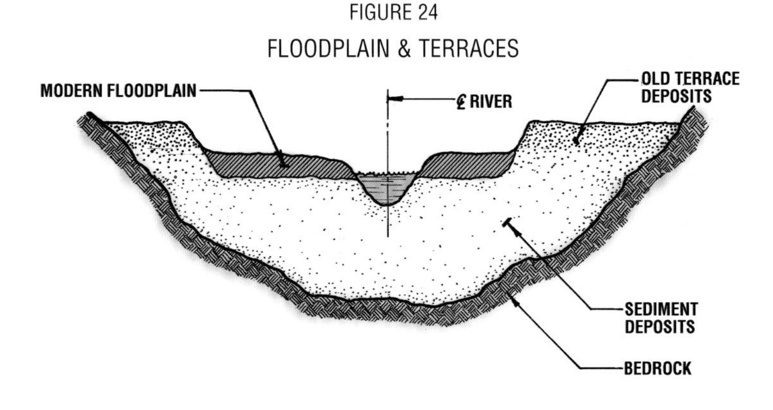

FIGURE 25

TYPICAL FLOW DEPTHS

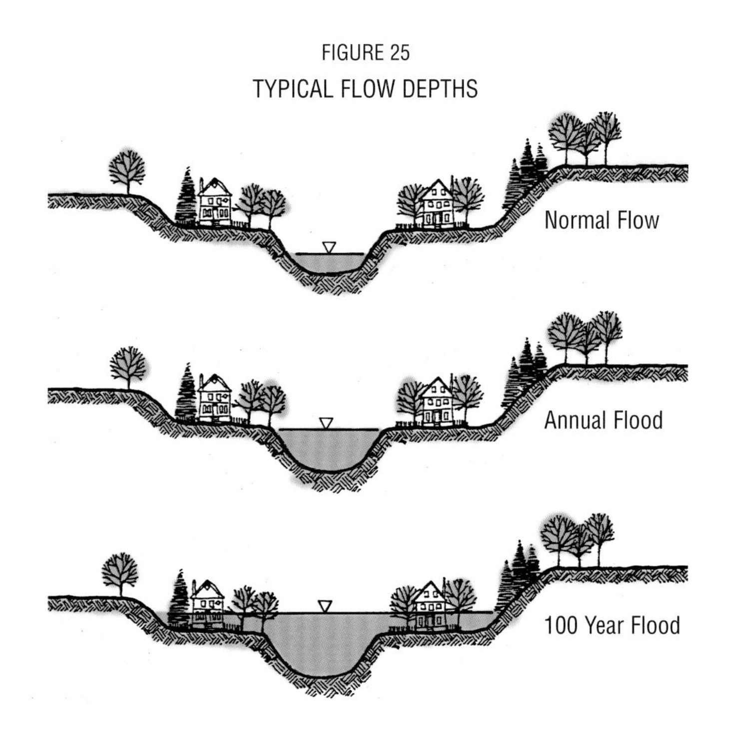

wood, or sediment from tributaries — fill with sediment. If a meander is cut off, it becomes an oxbow lake, eventually filling in during periods of flooding with fine grain sediments called clay plugs. (See Figure 22.)

The floodplain of the Farmington River in Simsbury exhibits many of the above features. For example, Pickerel Cove is an oxbow lake formed from an old meander that was cut off when a chute formed. (See photo on Page 47.)

The fresh, white sandy sediment shown in the photo on page 47 is a typical natural levee along a small river in Woodbridge. Such deposits gradually raise the bank higher than the floodplain. Eventually, the surface drainage of the floodplain is impeded, and backwater swamps may form.

Alluvial Terraces

Alluvial terraces are created when rivers erode a deeper channel, leaving their former floodplain at an elevation above current flood levels. Such transformation comes about as a result of changes in the Earth's crust, fluctuations in climate, or the impact of alterations caused by people.

These terraces are usually separated from the active floodplain by a steep scarp.

They may be paired, with one on either side of the valley at equal elevations. There also may be more than one level of terrace, representing formations of different ages. (See Figure 24.)

DEPTHS OF FLOW

Data indicate that the mean annual flow of a river is equaled or exceeded only about one-quarter of the time. It is thus a good indicator of flow conditions that can be considered normal.

Low flows form the deeper channel and the thalweg, which meanders across the bed. Minor floods tend to create and maintain the overall size of the channel. Major floods overtop the river banks and flow on the floodplain, shaping the floodplain in the process. (See Figure 25.)

Ordinary High Water Line

The ordinary high water line in a channel is the level along the bank that separates the predominantly aquatic area and the predominantly terrestrial area. It is evident from scour marks, soils, and changes in vegetation due to the prolonged presence of water. It will also roughly correspond to the mean winter ice level, since woody terrestrial vegetation can seldom survive ice movement.

Flow Capacity

The flow capacity and depths in alluvial rivers are related to the dominant peak flood size and frequency. This relationship may vary from one alluvial river to another.

Peak rates are expressed in terms of recurrence frequency. For instance, a 25-year-flood indicates a peak rate of flow that will occur statistically on an average of once in 25 years, with a 4 percent chance of occurring in any one year.

The mean annual flood discharge is the average flow of the largest annual flood in each of the years for which records have been kept.

Using data from 13 gauging stations in the Eastern half of the country, the United States Geological Survey has determined typical relationships for floods of various frequencies between depth of flow and height of channel banks, and between flow rate and channel capacity. (See Table 23.)

Flood Flow Depths

In Connecticut, the average flow depth of the 100-year flood has been used in the approximate mapping of flood-prone areas. The data, compiled by the United States Geological Survey, apply to free-flowing watercourses without obstructions such as bridges or dams, and are a function of the watershed size.

CROSS SECTION GEOMETRY

The channel cross section refers to its width, depth, perimeter length (the irregular line tracing the entire channel from the top of one bank to the other), and flow area (that part actually covered by water). Together, they describe the size and shape of a channel.

Width and depth may vary according to flow rate, slope, sediment load, and vegetation. High flow rates and steep channel slopes both lead to increased erosion and larger channels. Vegetation with dense roots tends to reduce erosion. Where a lot of sediment is carried by water, the river tends to be shallow and wider. If the sediment is suspended, the fine materials are deposited along banks during periods of low flow, encouraging growth of vegetation. A combination of cohe-

TABLE 23

TYPICAL FLOW DEPTHS

FLOW FREQUENCY	RATIO BETWEEN DEPTH OF FLOW AND BANK HEIGHT	RATIO BETWEEN FLOW RATE AND CHANNEL CAPACITY
Mean annual discharge	0.35	0.12
1.5 years	1.0	1.0
5.0 years	1.25	1.7
10.0 years	1.4	2.1
25.0 years	1.6	3.4
50.0 years	1.8	4.3

TABLE 24

DRAINAGE AREA AND FLOOD DEPTHS

Drainage Area in Square Miles	Range of Flood Depths in Feet *
2	3-8
10	4-10
30	6-12
70	7-14
150	11-16
300	15-20

* The values above should not be applied in the planning of individual development projects along rivers. Each such site has its own unique conditions.

FIGURE 26

CHANNEL CROSS SECTIONS

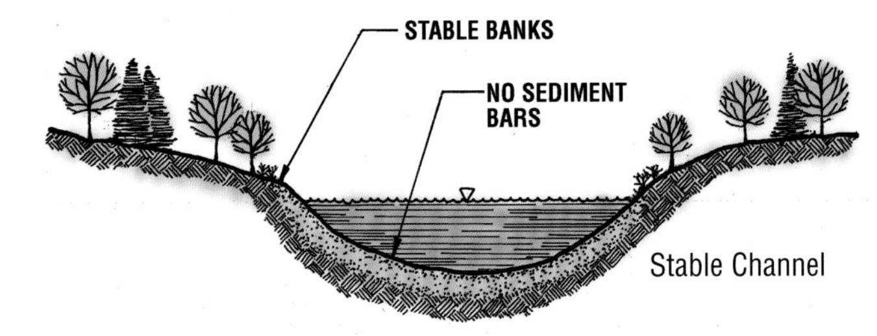

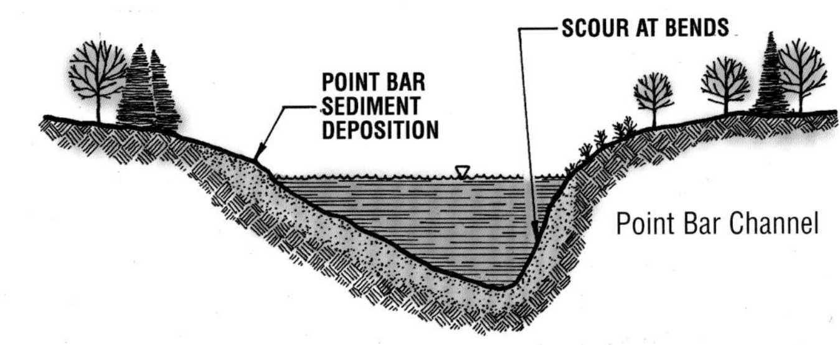

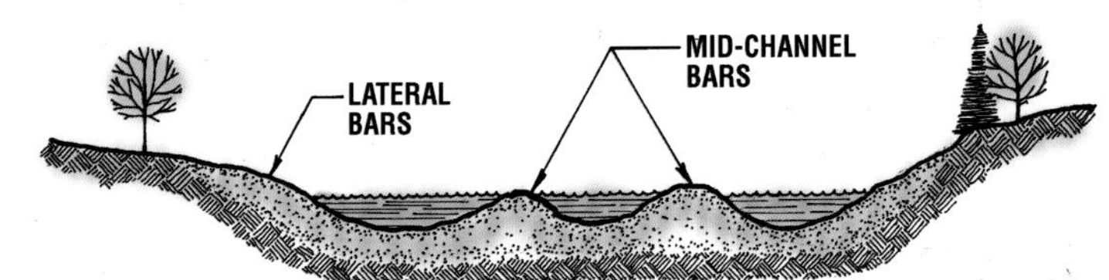

Braided Channel

A large point bar of coarse sand and gravel.

sive sediment and vegetation helps to stabilize the surface of banks and reduce their erosion.

Conditions that create abnormal channel size include steep, compact watersheds, a high percentage of impervious surfaces, long droughts with few channel-scouring floods, and artificial fill.

The following trends apply:

- Channel depth and width increase in size with increasing flow rates.
- The final channel size and shape is influenced by the resistance of the channel bed and banks to the forces of erosion.
- Channels in erosion-resistant material tend to be narrow and deep, while those in erodible soils are usually wide and shallow.
- Vegetation along the banks contributes to soil cohesion. This reduces erosion of the banks, but encourages erosion of the riverbed. Vegetated banks thus promote channels that are deep and narrow.
- Fluctuations in flow encourage wider channels with braids materializing during low flow periods. They change in form, shape, and size after each flood.
- In alluvial channels, changes that lead to equilibrium (long-term channel stability) prevail. Changes that do not lead to equilibrium are not enduring.
- The depth and width of an alluvial channel change when the amount or composition of sediment being transported changes.

Meandering channels tend to scour on the outside bank at bends. They deposit point bar sediments on the inside. As a result, triangular-shaped cross sections develop at bends. (See Figure 26.)

Braided rivers have excess loads of sediment. They develop mid-channel bars, splitting the water flow into a series of smaller channels.

Types of Channel Sediment Bars

There are many different types of sediment bars (See Figure 27).

Alternate bars are sediment deposits found along straight channels, generally against the banks, alternating from one side to the other. They force low flows to meander within the channel.

Point bars are sediment deposits that occur in zones of low-velocity flow at the inside of bends. They are found on the bank opposite to scour pools, which are on the outside of bends.

Junction bars are sediment bars where tributaries make deposits in the larger channels.

Mid-channel bars are crescent-shape deposits in the middle of a channel, between bends, generally forming riffles. They have a concave face on the downstream side, with points extended by fresh sediment.

Diamond bars are extended mid-channel bars, found in braided, sand- and graveled-bed channels.

Diagonal bars are found in gravel bed channels, forming riffles in mid-channel.

Sand waves are a series of dune-like deposits found in mid-channel, moving downstream in the shape of crescents.

Deltas

Deltas are a special type of deposit that occur where sediment-carrying rivers enter bodies of water much larger than the stream, such as lakes or the ocean. The resulting lower flow velocities allow sediment to settle, forming a delta deposit. Deltas may also form where tributaries enter a much larger river that has a low flow velocity.

Deltas can have any of several patterns, depending on relative velocities of flow in the river channel as compared with the receiving body of water.

Many lakes in Connecticut have deltas where streams enter them. Such deltas are rapidly covered with plants and grow slowly over the years.

Connecticut's rivers generally do not have deltas where they enter Long Island Sound. The larger rivers may have sediment bars at their mouths, but these marine bars do not grow rapidly enough to withstand the erosive force of tides and storm-driven waves, so they remain submerged.

Aggradation

Aggradation is the general increase in elevation of a long reach of a riverbed over a long period.

This process occurs when sediment is continually added to the riverbed, or even the floodplain, and the river does not have the necessary slope, velocity or flow rate to wash away the sediment. Therefore, the riverbed will rise, increasing the slope in relation to the segment farther downstream. This increased slope accelerates erosion, until sediment transport is equal to sediment supply rate and equilibrium is achieved.

FIGURE 27

CHANNEL SEDIMENT BARS

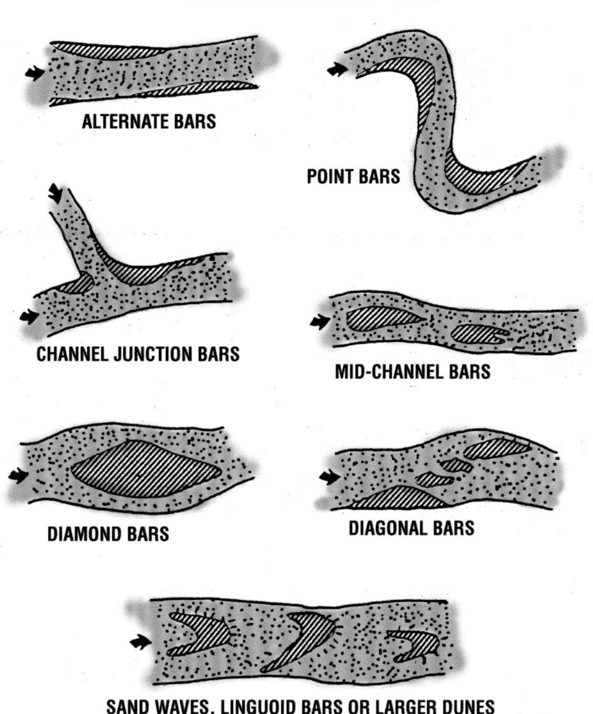

Bars that are stabilized with vegetation may become islands.

A delta extending from a road culvert into a pond, gradually filling the pond.

FIGURE 28

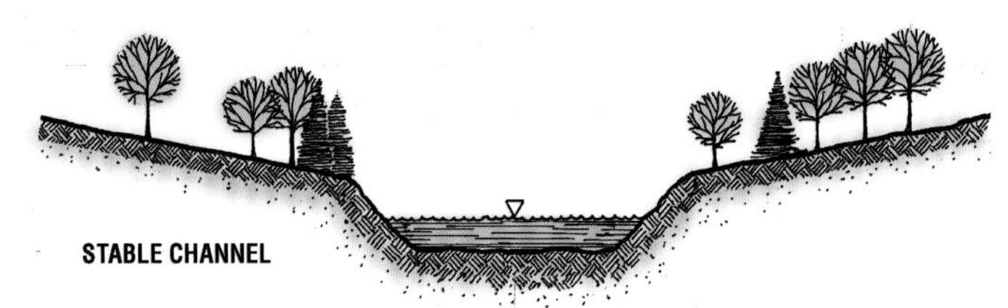

Channel bed elevation remains constant over a long period of time.

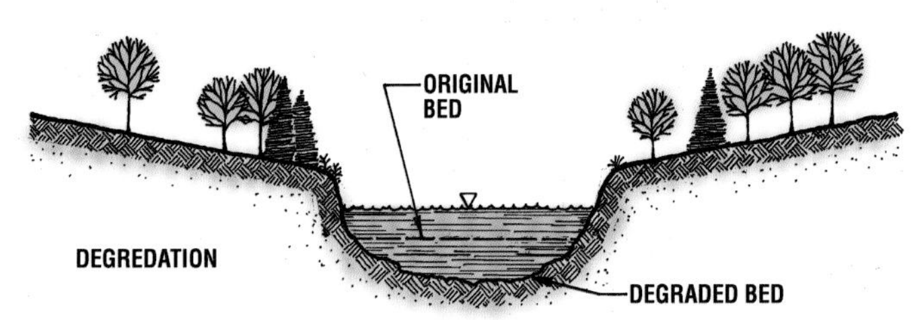

Channel beds that have long-term erosion to a lower elevation are degrading. The banks become steeper and may be unstable.

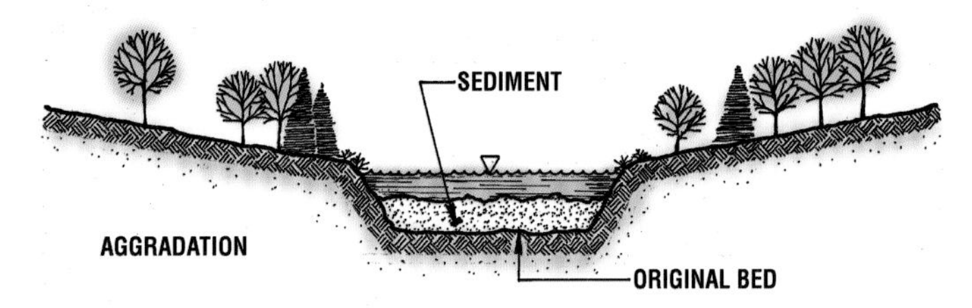

Sediment deposition fills in the channel, raising the channel bed and water profile elevations.

As the sediments are deposited downstream, the slope in relation to the upstream section is reduced, in turn reducing sediment transport. Thus the area of deposition migrates upstream. The sediments are segregated by weight into stratified layers, sloping downstream.

The aggrading channel often becomes braided as it fills. Overbank flows occur with greater frequency, leaving vertical accretion deposits on the surface of the floodplains. The floodplain surface may thus rise with the river. (See Figure 28.)

Among the factors that contribute to aggradation are a decrease in watershed vegetation and an increase in soil erosion, or a rise in sea level. These can be the result of climatic changes or of human activities such as logging, agriculture, and urbanization.

Aggradation is a somewhat unusual condition in inland Connecticut because the dense vegetative cover limits soil erosion in most watersheds.

Degradation

In contrast, degradation — the general lowering of the streambed — is common in Connecticut. This occurs where the slope, discharge and flow velocity combine to transport more sediment than is supplied to a river section. As a result, the riverbed will erode until the slope and velocity are reduced to a point of equilibrium.

Degradation can occur as a uniform lowering of the bed. In other cases, where two stable portions of the river are separated by a steep reach, degradation can occur as the bases of scarps or headcuts in the channel erode, causing them to migrate upstream.

Natural degradation can result from an uplift of the land, climatic changes, or even an increase in vegetation. Humans can cause or accelerate degradation through watershed development that increases surface runoff and flow rates. Dams on alluvial rivers encourage degradation by trapping sediment that would normally be carried downstream.

Degradation continues until equilibrium is reached. An exception is in streams with a mixed-grain bed. In such cases, the river tends to erode the fine-grained sediments first, leaving heavier, coarser material behind. If the latter is large enough, it will resist erosion and slowly accumulate, forming an armored bed, which prevents further degradation. The river may then erode laterally, or it may reduce the size of the coarse particles by abrasion until they are eventually transported away.

ENTRENCHED CHANNELS

An entrenched channel is one that has degraded so much that its flood flow is unable to spread across its floodplain. Such channels are confined by well-defined banks that are higher than the mean annual flood level, thereby preventing inundation. Entrenched meanders occur when the channel's original pattern was preserved as the channel degraded, such as in the Grand Canyon. In other words, entrenched meanders are those that have eroded vertically but not laterally. They have steep valley walls on both sides of the meander bends.

Incised meanders occur where the channel has eroded both vertically and laterally. They move downstream by eroding the outside of the bends. They are characterized by steep valley walls on the outside of bends, with mild sloping walls on the inside. The Scantic River in Enfield and South Windsor has a classic entrenched floodplain with an active meandering channel. The condition exists because the Scantic flows through the highly erodible sediments of glacial Lake Hitchcock, which formed in the Connecticut River Valley during the glacial melt period.

A small underfit alluvial stream meandering across a relatively broad floodplain.

MISFIT STREAMS

A stream whose natural meander lengths do not match its valley meander lengths is said to be a "misfit." Such streams exist where earlier flow conditions were vastly different from those of the present.

For example, runoff from melting glaciers may have carved a valley much larger than could be formed by the relatively lower flows of today. Thus, misfit streams occupy only a portion of the valley bottom. Streams are called "underfit" when their meander lengths are smaller than those of the valley.

Such valleys have large terraces corresponding to the earlier floodplain level that are a record of the former meander sizes and discharge rates. (The repeating pattern of curve radius and curve length distinguishes valley meanders cut by the river from irregular winding valleys.)

Research shows that the meander lengths of selected bedrock valleys in southern New England appear to have been carved by rivers much larger than those that exist today. (See example, Figure 29.) The five-to-one ratio of valley meander to stream meander suggests the past flows were 20 times as large as present ones.

In other parts of the country, the ratio of the valley meander to stream meander varies from three-to-one in the Ozarks, which were

FIGURE 29

SCANTIC RIVER INCISED FLOODPLAIN

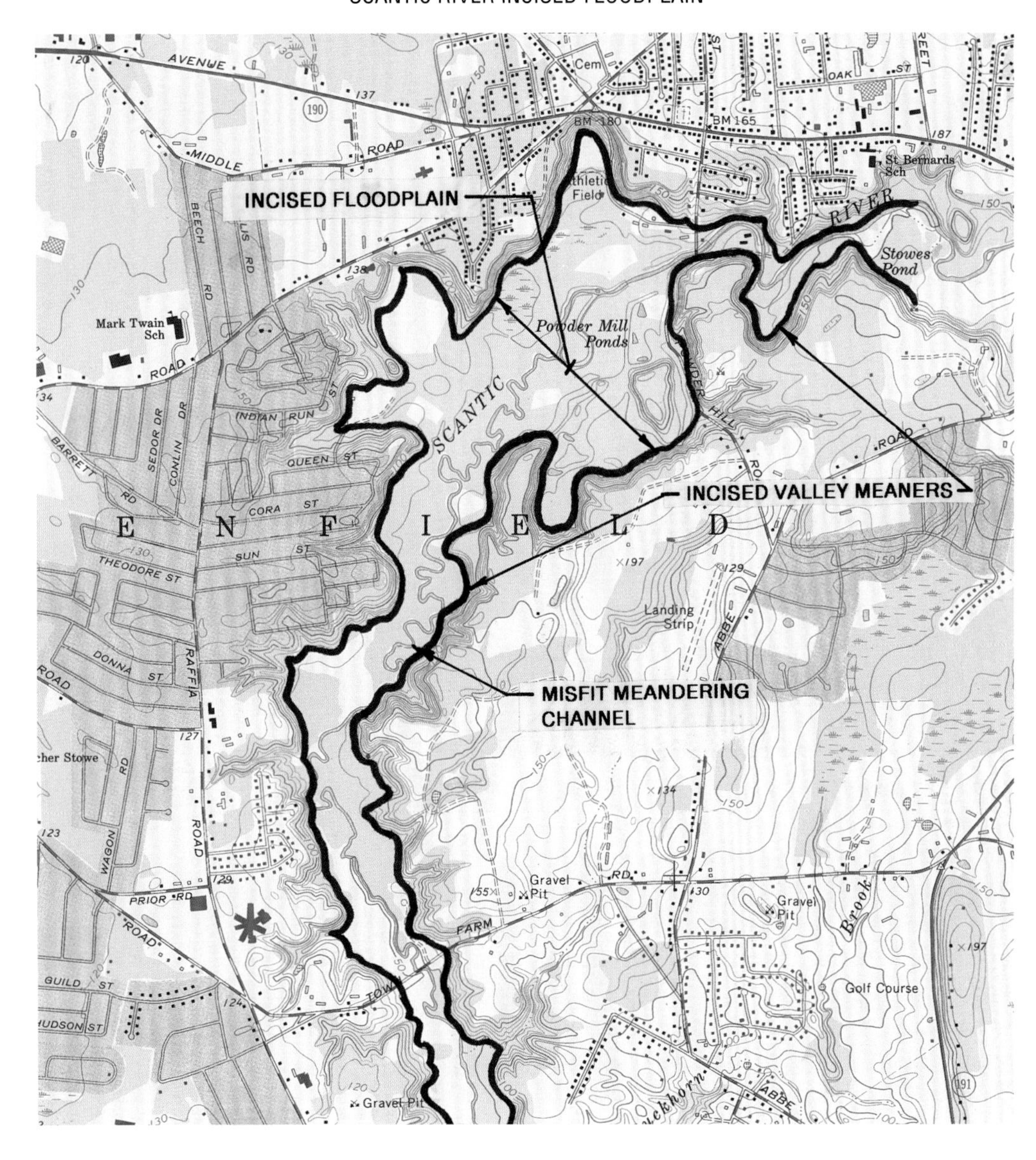

not glaciated, to 10-to-one in regions of Wisconsin that were subject to glaciation.

GROUNDWATER AND STREAMFLOW

Most rivers receive water not only from runoff, but also from groundwater sources. Streams may also recharge groundwater, particularly where they flow over permeable, stratified drift deposits of sand and gravel.

Influent streams are those whose flow infiltrates into the soil. This occurs where the water level in the river is higher than the groundwater level. Such is the situation where upland streams flow down over stratified drift deposits.

In many cases, the yield of an aquifer depends on the rate at which rivers recharge underlying sediments. When wells reduce the elevation of groundwater, they can encourage more infiltration, thereby reducing streamflow. This is called induced infiltration.

Effluent streams are those that receive discharges of groundwater. This commonly occurs in upland areas. During dry weather, streamflow may be maintained by water draining out of the soil and rock.

An individual stream may be influent or effluent, depending on the season. For example, streams receiving sufficient flow from their upland areas tend to recharge groundwater during dry weather. During wet weather, when groundwater is high, it seeps back into streams. (See Figure 30.) This alternating sequence helps to maintain base flow rates between rainstorms.

An ephemeral stream does not have a base flow, as the channel is always above groundwater.

FIGURE 30

EFFLUENT AND INFLUENT CHANNELS

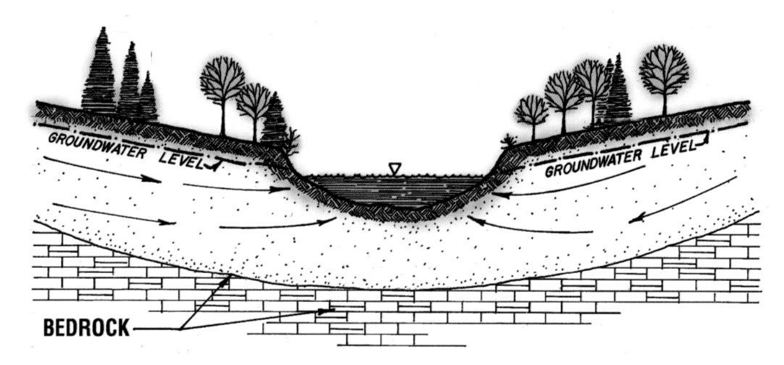

Groundwater flows to the stream, increasing the streamflow. – **EFFLUENT**

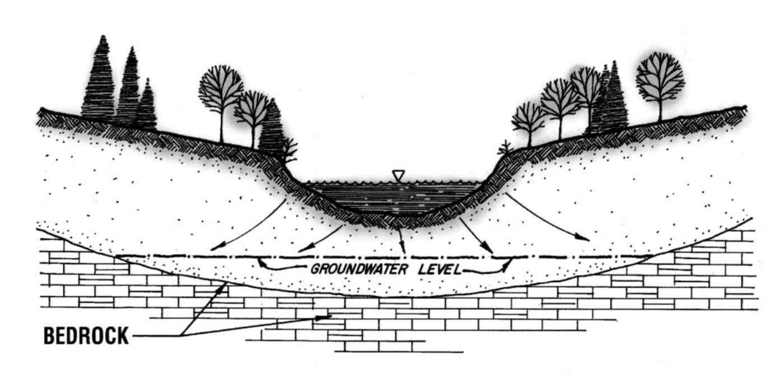

Groundwater levels are below the stream level. Streamflow infiltrates into the soil to the groundwater. – **INFLUENT**

RIVER ICE

Most rivers in the northern United States are subject to freezing in the winter.

During a normal winter in Connecticut, streams typically develop an ice cover of between one and 12 inches thick in all but the fastest-flowing segments.

River ice can affect the slope and vegetation of the stream's banks, and can form jams that have the potential to cause flooding.

Ice can form in several ways. Layer ice forms on slow-moving or stationary water, then thickens by growing downward as new horizontal layers are added to its base. Often referred to as sheet ice or

Ice jams may obstruct streamflow and lead to flooding.

black ice, this variety, as it thickens, has a strong, smooth, uniform surface.

Skim ice, in contrast, has a vertical rather than horizontal structure. It tends to be white, and it cracks easily. It forms on ponds and on rivers with low velocity (less than two feet per second) when it is extremely cold and there is some wind or wave action. The vertical structure results from the adhesion of loose pieces of layer ice broken up by the wind, waves, or current. The open spaces between the skim ice fragments fill with weak frazil ice subject to subsequent cracking.

Border ice forms where water comes into contact with rocks, vegetation, or other bank material that is extra cold. Such ice gradually grows outward, increasing in width, with an irregular surface. On small streams, it may extend from bank to bank, meeting in the center.

Anchor ice forms on cold submerged surfaces and grows upward from the streambed.

Frazil ice originates in open, fast-flowing water as crystals form on and around suspended particles. The individual particles of ice grow as they are carried by the current, eventually adhering, when they may freeze into slush and form larger blocks. Frazil ice can accumulate, sometimes causing dams and jams that obstruct river flow. Since frazil ice can form only in open water, even a thin layer of surface ice can protect against frazil ice conditions. Frazil ice usually is white or opaque.

Once ice forms on a stream, snow often builds up on its surface in a series of strata that can become compacted into weak snow ice.

Ice can contribute to flooding. Floating ice can increase friction along the surface of flowing water, reducing its velocity. Because this surface friction impedes drainage and because ice takes up space in the channel, an ice-covered stream has a higher water level for a given flow than it does when it is not frozen. High-velocity flow rates may cause ice to fragment and be carried downstream.

The ice thickness is reduced during thaws by surface melting on the top and by the turbulent water eroding the bottom surface. Increasing streamflows lift the floating layer ice, creating tension cracks and breaking the ice mass into jagged pieces.

Jams occur when loose frazil ice or fragmented sheet ice floats downstream and becomes concentrated, blocking part of the channel. The broken ice also accumulates at constrictions or bends in the channel or when it encounters obstacles such as bridges or islands. Two Connecticut locations subject to chronic ice jams are the

Housatonic River in New Milford and the Salmon River in East Haddam, where an ice jam in 1988 began at a bridge.

Floating jams occur when fragmented ice accumulates over flowing water. More serious bottom jams occur when upstream ice overrides downstream ice, forcing it deeper into the channel. This can result in flooding upstream from the blockage, or downstream flooding when the jam breaks up and releases large volumes of water.

Ice exerts great pressure on the banks and bed of a river as well as on man-made structures. Ice that has frozen along the bank can contribute to erosion when it breaks loose, carrying attached earth, rocks, and vegetation downstream.

The presence of moving ice limits the growth of woody vegetation near the water line. The ice that freezes along the river bank will adhere to trees and shrubs and tear them out of the ground when the ice moves. Consequently, only emergent and flexible-stem vegetation is found below the average elevation of winter ice on river banks.

On larger rivers, ice can exert tremendous horizontal and vertical pressure on piles that support docks, piers, and other shoreline structures, causing damage that necessitates annual maintenance. Ice that forms around piles can literally pull them out of the ground when water levels rise. Ice can also freeze around protruding pieces of rock riprap and move it out of position.

The regular dimensions of this canal mark it clearly as a man-made channel.

CANALS

In the 19th century, population growth and industrial development encouraged an era of canal building. Connecticut still has many miles of canals, most of which were abandoned years ago. Besides being of significant historical interest, these canals — some of which have been converted to flowing watercourses or have reverted on their own — form a unique type of channel.

Canals were used to supply water and to power mills. (Some are still used for such purposes.) Barge canals were constructed in valleys to provide level waterborne transportation where rivers were unnavigable. Drawn by horses or mules, the barges carried products to market, occasionally transporting passengers as well.

Typically, a canal was 12 to 20 feet wide and four feet deep. Vertical-lift locks were employed to overcome changes in elevation. Towpaths were located along the banks. (See Figure 31.)

The Farmington Canal, an 80-mile-long corridor from New

Many canals were built to transport goods around rapids on otherwise navigable rivers.

FIGURE 31

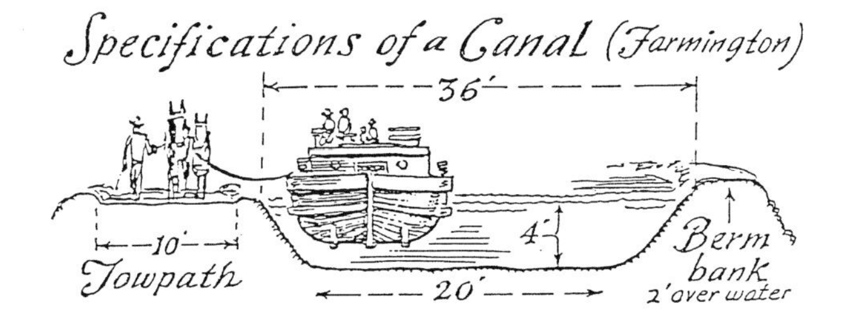

Haven to Northampton, Mass., was completed in 1836. It followed the natural watercourses of the central valley, Shepard Brook, Mill River, Ten Mile River, the Quinnipiac River, and the Farmington River. A series of 28 locks was required to adjust its grade over a 220-foot ascent. The original canal had a minimum width of 20 feet and was four feet deep. It handled barges up to 74 feet long and weights of 25 tons. However, the canal was plagued by floods that washed out its berms. Later, droughts left it with insufficient water.

The Windsor Locks Canal was built in 1829 to allow barges to bypass the Enfield rapids, which until then was the upstream limit of maritime navigation on the Connecticut River. Early cargoes had to be unloaded, carried around the rapids, and then reloaded onto smaller barges for transportation farther upstream. The 5.5-mile canal allowed the river to be a key transportation corridor well into the 20th century.

Some of these old canals now function as part of the state's drainage system. However, they have severe limitations as watercourses.

First, their flatness leads to low flow and aggradation. Portions of Willow Brook in Cheshire, for example, coincide with the Farmington Canal but do not have slopes sufficient to handle flood runoff and sediment, thereby resulting in periodic flooding. Another problem has arisen in Hamden, where the same canal has "captured" Shepard Brook. In this case, the raised towpath has become a dike, separating runoff from the floodplain and preventing floodwaters from reaching the channel and draining away.

Water flows over a series of drops in a bedrock channel.

TABLE 25

SUMMARY OF SIGNIFICANT FEATURES OF UPPER, INTERMEDIATE, AND LOWER STREAM TYPES.

UPPER (YOUTHFUL)	INTERMEDIATE (MATURE)	LOWER (OLD AGE)
GENERALLY DETRITUS-BASED ECOSYSTEM	***GENERALLY PHOTOSYNTHESIS-BASED ECOSYSTEM***	***GENERALLY SEDIMENT-BASED ECOSYSTEM***
• High gradient (more than 20 feet per mile); water velocity is fast • Characteristic of headwaters and upland mountain reaches with steeply sloping banks • Water flows throughout the year but can be highly seasonal, flashy • Substrate consists of rocks, cobbles, or gravel with occasional patches of sand • Little floodplain development; typically flows through narrow valleys • Natural dissolved-oxygen concentration is normally near saturation • Fauna is characteristic of running water; there are few or no planktonic forms; burrowers may be found in silt or sandy pools, but majority of organisms are "torrential" water forms that cling to rocks or inhabit interstitial areas • Generally cold water	• Moderate gradient (10 to 20 feet per mile) • Characterized by steep, fast-water stretches interspersed with deeper pools in which the current is slowed • Characteristic of upper valleys, foothills, and intermediate-width floodplains • Streamflow fluctuates seasonally • Diverse biologic habitat supporting a great variety of stream life	• Low gradient (less than 10 feet per mile); water velocity is slow • Water flows throughout the year and is relatively constant, supported even in dry years by minimum base flow • Substrate consists mainly of sand, mud, and silts • Floodplain is characterized by broad, flat valleys • Dissolved-oxygen deficits may sometimes occur • Fauna is composed mostly of organisms that reach their maximum abundance in still water and includes organisms that burrow into the silt and sand • Generally warm water

Source: "Manual of Stream Channelization Impacts on Fish and Wildlife," Fish & Wildlife Service, US Dept of the Interior. Kearneysville, West Virginia 1982

CHAPTER THREE Stream Ecology

What would the world be, once bereft
of wet and of wildness? Let them be left,
O let them be left, wildness and wet;
Long live the weeds and the wilderness yet.

Gerard Manley Hopkins, 1844-1889

FIGURE 32

STREAM ECOLOGY

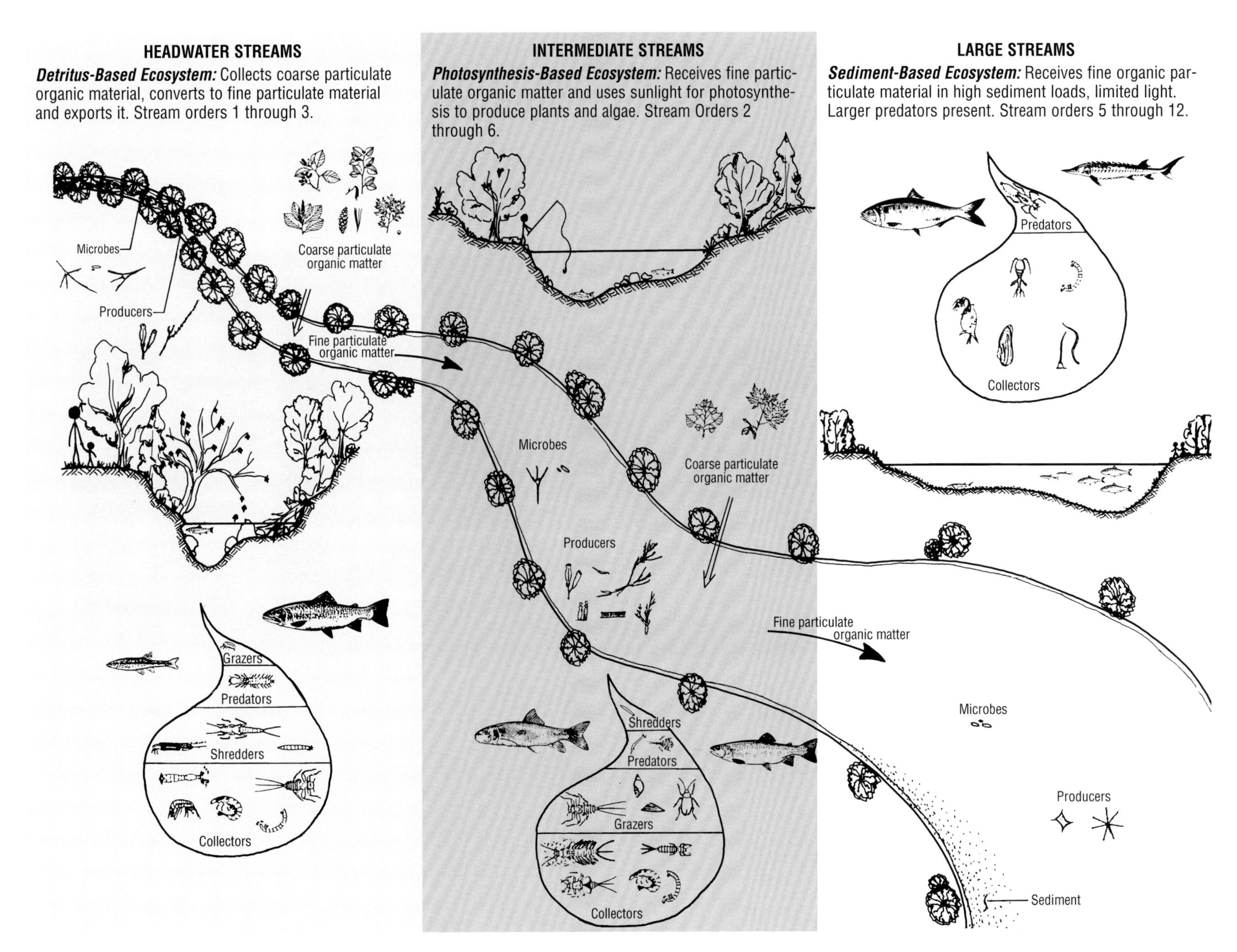

Stream Ecology

River corridors provide several distinct habitats to which plants and animals have adapted. Each type of habitat meets the needs of selected species of plants and animals, providing suitable terrain, space, shelter, food, and water.

Plants are often called "producers" because, through photosynthesis, they use the sun's energy to create their own food from basic nutrients and inorganic matter, such as carbon dioxide and water.

Animals depend for their food on organic material from plants or other animals, and so are often called "consumers." Herbivores are animals that eat plants; carnivores eat other animals; and omnivores eat both plants and animals. Scavengers are omnivores that eat plant and animal remains, thus recycling the organic material.

Windblown tree trunks, brush, and irregular banks trap organic material that is essential in the aquatic food chain.

HABITAT ZONES

River habitats can be subdivided into benthic, aquatic, and terrestrial zones.

The benthic zone consists of the streambed and the plants and animals that live in, under, or close to it. In this zone, species generally are attached to or buried in the substrate and are accustomed to being submerged.

The aquatic zone includes the flowing water and the animals in it, such as fish, insects, reptiles, amphibians, and some mammals. It also includes floating plants such as algae.

The terrestrial zone comprises the adjacent upland and the plants and animals that live on land that is seldom submerged. This zone favors animals that visit both land and water.

River habitats also vary with local conditions. For example, many of the plants and animals in a shaded, swift-flowing mountain stream will be different than those associated with a deep, broad river with warm, slow-moving water.

TABLE 26

BED-MATERIAL-BASED STREAM REACH CLASSIFICATION

Bed type	Particle size (mm)	Relative frequency of bed movement	Typical benthic density	Macroinvertebrate diversity	Fish use of bed sediments
Boulder-cobble	>64	Rare	High	High	Cover, spawning, feeding
Cobble-gravel	2-256	Infrequent	Mod.	Mod.	Spawning, feeding
Sand	0.062-2	Frequent	High	Low	Off-channel fine deposits used for feeding
Fine material	<0.062	Varies	High	Low	Feeding

Source: "Sediment and Aquatic Habitats in River Systems," American Society of Civil Engineers Journal of the Hydraulics Division, May 1992

TABLE 27

Life Processes and Channel Characteristics

CHANNEL CHARACTERISTIC	DETRITUS ZONE	PHOTOSYNTHESIS ZONE	SEDIMENT ZONE
Energy source	external	internal	imported
Typical channel width	0-15'	15-100'	over 100'
Solar exposure	low	high	moderate
Channel roughness	high	variable	low
Channel slope	variable	variable	low
Terrestrial vegetation dominance	high	moderate	low

Notes:
The detritus zone is usually in small, upland streams with a continuous vegetated canopy.
The photosynthesis zone occurs in small to medium-size streams that receive solar radiation, which stimulates the growth of algae and other plants.
The sediment zone is characterized by high turbidity and sediment loads that limit light penetration and photosynthesis. It is usually found in larger rivers.

Moss and algae — here growing on a stable, rocky channel bed — are typical of streams with photosynthesis.

Examples of the variety of localized habitats associated with rivers and streams include pools, runs and riffles, as well as variations in the river's bank and bed, which may or may not be vegetated and may consist of either coarse or fine sediments or of bedrock. (See Table 28.)

River Ecosystems

River ecosystems vary along the length of a channel as it grows from a small rivulet to a large river. There may be considerable overlap. However, there are three basic ecosystems corresponding to the different ecological processes in small, intermediate and large channels. Each ecosystem can be further subdivided into slow and swift-flowing, fine or coarse-grained substrate habitats. (See Table 25.)

The three ecosystems correspond to three basic, sequential ways in which the biota are fueled with life-sustaining energy: from detritus, photosynthesis, and sediment. Each in turn becomes the foundation of the food chain as the geomorphological and hydrological characteristics of a river change as it flows from the headwaters to the mouth.

In the first of these zones, a stream receives organic material directly from the adjacent landscape through leaf-fall and woody debris. In the second, organic material is also produced in the stream by the growth of plants. And in the third, the river receives most of its organic material in the form of sediment from sources upstream and direct land runoff. (See Figure 32 and Table 26.)

This "river continuum concept" helps to explain the biological processes in streams: where organic matter comes from, how it moves, how it is stored, and how it is consumed by the ecological communities.

A river is a complex, open system, a turbulent environment, always in flux. In contrast, lakes and ponds are generally closed systems where the materials necessary for life are continually recycled. In rivers, nutrients pass by only once as they are transported downstream. Many aquatic species are totally dependent on what happens to drift by at a given time.

In such a complex system, human activities can have severe consequences. For example, pollution introduced into a river may be carried far downstream, adversely affecting any animal life that comes into contact with it. Although some toxic materials may be diluted as they flow, others can accumulate in the streambed, where their effects may continue for a long time.

Detritus-based Ecosystems (Small Stream Systems)

Small streams, of orders one through three, usually have channels less than 15 feet wide. In Connecticut, under natural conditions, these channels often have continuous foliage canopy that limits sunlight. Small streams interact with the adjacent terrestrial systems, collecting, converting, and transporting nutrient material. Such streams usually have high levels of dissolved oxygen, due to groundwater discharges, cool water temperatures and turbulence. The flow rates and depths vary during storms because of the large volume of direct runoff and limited floodplain storage.

Most of the initial energy source in this ecosystem is in the form of leaves, twigs and wood particles that fall into the water. This coarse particulate organic material is generally not mobile. It tends to be trapped along the bed, banks and rocks, amid logs, branches, and roots. These obstructions are important, since they retain the material before it is carried downstream, facilitating partial decomposition by bacteria and fungi. Several types of aquatic insects also feed on this detritus. This partially decomposed material is then carried downstream in the form of fine particulate organic matter that is later consumed in the next ecological zone.

Fish in small streams feed primarily on insects, such as mayflies, caddisflies, and stoneflies. Fish species include minnows, suckers, darters, trout, and sculpins. Small streams are sensitive to periods of low flow. Occasionally, they have almost no flow, forcing resident fish into small pockets or pools where there are limited amounts of oxygen. These pools are important refuge areas that help fish survive during droughts.

Forest detritus is the initial basis of the stream's organic load and food chain. Any activity that alters the detritus sources in headwater streams will affect the downstream ecosystems. Thus, it is important to preserve riparian vegetation, as well as to avoid making any changes to the snags and leaf jams in the channel that trap organic material. Streams with unstable banks or smooth, uniform channels accumulate less such material. Small streams in non-forested landscapes receive less organic material than those that flow through woodlands. The former also are subject to large fluctuations in water temperatures because of their greater exposure to solar radiation. In these situations, a small stream may function more like a warm-water, intermediate-sized one. High water temperatures can reduce dissolved-oxygen levels, contributing to fish kills, especially in shallow water.

TABLE 28

THE FOOD CHAIN

Detritus	Natural debris such as leaves and twigs that falls into streams from the adjacent upland. Detritus is generally the first organic contribution to the aquatic food chain. Leaves and wood that arc trapped or snagged serve as a primary food source for many fungi, bacteria and insects.
Fungi and Bacteria	Large groups of microorganisms that lack chlorophyll and live in colonies on the leaves and twigs that fall into the water. Fungi and bacteria decompose the material, softening it, reducing its mass.
Phytoplankton	Algae and other plants, usually microscopic, that are free-floating in the water column. The abundance of suspended algae is common in larger rivers whose waters are relatively clear and where sunlight is abundant. When present, algae provide a significant nutrient source in intermediate rivers and streams.
Zooplankton	Free-floating or weak-swimming animals, usually microscopic. Principal examples include crustacean and fish larvae.
Macrophytes	Larger vascular plants that grow in water. Often they provide a substrate for attached algae, bacteria, and fungi. Macrophytes include many aquatic flowering plants, mosses, and larger algae.
Shredders	A group of insects that feed on bacteria, algae, and smaller invertebrates that are attached to detritus in the water. In the process, the leaves are partially consumed and shredded, releasing smaller particles of organics and fecal material, which move downstream. Species include some cranefly, caddisfly, and stonefly larvae.
Collectors	These are insects that feed on bacteria and other fine particulate organic material created by shredders. Collectors obtain the material by filtering it from the water and from the bed. Species include some black fly, caddisfly, and midge larvae.
Scrapers and Grazers	These are invertebrates that obtain food by scraping algae and moss from the submerged streambed and banks. Algae and mosses survive in well-lighted streams through photosynthesis. Scraper and grazer species include some caddisfly and stonefly larvae.
Predators	These are invertebrates and fish that feed on shredders, collectors, scrapers, and other animals. Cold-water fish species are common in shaded and swift streams, while larger numbers of bottom feeders are found in larger, warmer rivers.

A beaver dam can dramatically alter the nature of a stream, creating habitat for a variety of wildlife.

Fly-fishing on the Salmon River.

Photosynthesis-based Ecosystems (Intermediate Stream Systems)

As small streams flow downhill and join other tributary watercourses, they gradually grow to intermediate size, orders two through six, with channel widths of 15 to 50 feet. As the channel widens, the canopy shades a smaller proportion of the water and the stream receives more sunlight. Water temperatures are warmer, and there is increased photosynthesis by algae, mosses and vascular plants attached to the banks. Through this process, plants produce an excess of nutrient materials, thereby creating a net gain in the stream's biological level.

Intermediate systems receive residual fine particulate organic matter from upstream reaches and a moderate amount of direct coarse material such as leaves and twigs. Floods help to scour, resuspend, and transport such material, which accumulates during periods of low flow. The invertebrate community dominated by scrapers, grazers and collectors lives in such a setting.

Due to the variety of food and habitats, intermediate streams are home to a large number of species. Biological activity is high. Fish populations are diverse, varying with the morphology of the channel.

Sediment-based Ecosystems (Large River Systems)

Larger streams, orders five through 12, are usually more than 50 feet wide. They receive most of their nutrients from a combination of input from tributaries, internal photosynthesis and the recycling of detritus. The nutrient input from bank vegetation is only a small percentage of the total.

Mature, large rivers usually have low velocities and limited turbulence. Although they have the potential for high photosynthesis, the murkiness of the sediment- and algae-laden water may block the penetration of sunlight and limit the abundance of macrophytes. Because the demand for oxygen by bacteria and invertebrates often exceeds that which is produced by plant photosynthesis, depressed oxygen levels are common. Many such rivers receive discharges of sewage, which may deplete oxygen levels further, both directly as it decomposes and indirectly by stimulating excessive growth of algae, whose massive die-offs consume oxygen.

Large rivers shelter plankton, invertebrates, and fish, large and small, including both predator and foraging species.

FISH HABITAT

A river is a harsh environment. Fish must adapt to a variety of difficult conditions to survive. It is no accident that the fish in fast-moving rivers have high stamina and streamlined bodies. To maintain position in a favored region, they must continually combat the current. Many seek deeper, less turbulent pools. Yet even here the forces of the current operate.

The composition of fish species in a given watercourse depends on a number of important factors. Glacial history and human introduction are responsible for the present-day fish fauna found in Connecticut's different drainage basins. Water temperature, water quality, food sources, dissolved-oxygen levels, and the physical characteristics of the channel determine the presence and abundance of species.

Adequate water depths are necessary to allow fish to swim and provide cover from predators and sunlight. Fish "hide" in dark shadows, which offer protection while allowing them to dart out to feed on insects and smaller fish. Greater depths increase the size of the habitat and the number and size of fish it can support.

The flow velocity and characteristics of the streambed substrate are closely related. High velocities are associated with gravel and cobble substrates, and with greater turbulence. Low velocities are associated with fine grain, silty substrates.

Each species has its own preference as to where to feed and spawn. Specific species of fish prefer to feed on specific species of insects and plankton, each of which thrives in a limited range of substrate types.

Trout and salmon lay eggs in coarse gravel bars of swift-flowing streams. These bars provide a stable, sheltered location for the eggs, while cool water flowing through the voids between particles supplies them with oxygen. Burying coarse gravel beds with sediment reduces reproduction of trout and many other species. As gravel beds become embedded with sediment, production of aquatic insects, a food source on which fish depend, is also reduced.

Boulders, undercut banks, logs, and pools provide protection from predators and locations for feeding. Typically the more cover a stream contains the greater the number of fish it will support.

To provide a complete habitat, areas for feeding, spawning and resting must be interconnected, allowing fish to move from one to another.

Most fish are very sensitive to changes in water temperature and

TABLE 29

KEY CHANNEL CHARACTERISTICS THAT DETERMINE THE ABUNDANCE OF SPECIES

- Depth of water
- Volume of water
- Velocity of flow
- Composition of substrate
- Cover and shelter
- Continuity of habitat

TABLE 30

TEMPERATURE REQUIREMENTS FOR SELECTED FISH SPECIES

(DEGREES FAHRENHEIT)

SPECIES	OPTIMUM TEMPERATURE	CRITICAL LIMITS
Pike, pickerel	48-85	35-90
Catfish, bullheads	70-90	32-95
White perch	50-82	32-88
Striped bass	50-82	32-88
Trout	50-68	32-76
Suckers	50-68	32-92
Bass, sunfish	56-87	35-92
Herring, shad	52-72	32-76
Minnows	46-84	32-95
Carp, goldfish	60-90	32-100

TABLE 31

HOW HABITATS ARE DAMAGED

- Building artificial channels
- Erecting dams or other barriers
- Clearing banks or removing tree canopy
- Altering substrate
- Disturbing rooted plants
- Eliminating instream shelters
- Reducing flow
- Increasing flow velocity
- Disturbing spawning areas
- Reducing surface area of water
- Destroying pools and riffles

An old industrial hydropower dam obstructs fish movement.

dissolved-oxygen levels. Very cold temperatures reduce fish activity and feeding, while warm temperatures can reduce oxygen levels and cause fish kills.

Fish Habitat Degradation

Physical changes in a river can affect a species by altering its habitat. (See Table 31.) Many streams in Connecticut have been obstructed by dams that prevent free migration. Others have been artificially lined with concrete or unnaturally diverted. Dredging, channelization and other human activities affect flow rates and alter the channel bottom, not only at the site of the disruption, but also farther downstream as a result of increased erosion and sedimentation. Removal of streambank vegetation is especially harmful. Such activities can destroy a habitat. Even changes to pools and riffles have repercussions.

The introduction of exotic fish also can harm native species. Non-native fish may eat native species or the food sources they depend on. In addition, exotics may not be affected by the natural checks that keep fish populations in balance, or may carry diseases to which endemic species have not developed immunity.

There are many techniques available to protect aquatic habitat. (See Table 32.)

Recreational Fishing

While commercial fishing in Connecticut is limited to coastal areas and some of the larger rivers, the state's inland waterways are used extensively for recreational fishing.

Some of the state's popular game species are: trout, throughout the state; smallmouth bass, in large streams and rivers; largemouth bass and catfish, in large rivers; panfish, in many intermediate and larger rivers; herring, white perch and American shad, in coastal rivers and tributaries of the Connecticut; and striped bass, in lower reaches of large coastal rivers.

The Fisheries Division of the Connecticut Department of Environmental Protection manages the recreational fisheries and works to improve aquatic habitats, to restore native species such as Atlantic salmon, and to prevent introduction of destructive exotic species or control those already introduced.

The DEP also stocks more than a half-million brook, brown and rainbow trout into the state's streams each year.

The United States Fish and Wildlife Service estimates that more than 1.9 million fishing trips are made by trout anglers in the state annually. 1.2 million trips are made to rivers and streams. Freshwater fishing generates an estimated $63 million a year in net economic activity.

It is also important to note that some people have to fish for food out of economic necessity. This is especially true in urban areas, where the rivers often also serve as waste receivers and water quality can be poor. This underscores the importance of safeguarding the ecological health of Connecticut's waterways.

INLAND WETLANDS

Wetlands occur where there is sufficient water to saturate the soil for long enough periods so that it can support certain species of animals and plants specifically adapted to those conditions. Wetlands play a critical part in the hydrologic system and have important ecological functions. Wetlands are an integral part of the river ecosystem.

Scientists have devised several ways to classify wetlands. In 1979, the United States Fish and Wildlife Service established a system, used by federal regulatory agencies, that takes into account vegetation, soil type and hydrology. It defines wetlands as a transition between terrestrial and aquatic systems where the water table is usually at or near the surface or the land is covered by shallow water. Wetlands must have one or more of the following attributes: at least periodically, the land supports predominately hydrophytes (plants that grow in water or on a substrate that is at least periodically deficient in oxygen as a result of excessive water content); the substrate is predominantly undrained hydric soil; or the soil is saturated with water or covered by shallow water at some time during the growing season each year.

The state of Connecticut defines inland wetlands as land consisting of any soil type that is poorly drained, very poorly drained, alluvial, or floodplain. Watercourses are defined as rivers and smaller streams, as well as lakes, ponds, marshes, swamps, bogs, and all other bodies of water, natural or artificial, vernal or intermittent, public or private, in or flowing through the state. Wetland soils in Connecticut are defined by soil drainage class and landscape position.

Poorly drained soils occur where the water table is at or just below the ground surface, usually from late fall to early spring. The land where poorly drained soils occur is nearly level or gently sloping. Many of our red maple swamps are on these soils.

TABLE 32

MEASURES THAT PROTECT AN AQUATIC HABITAT

Maintain vegetated buffer zones	Buffer zones help trap sediment. They shade water, minimize water-temperature rise, filter runoff, and maintain organic input to stream. The Connecticut Department of Environmental Protection Inland Fisheries Division recommends 100-foot buffers along perennial streams, and 50-foot buffers along intermittent ones.
Provide separation distance between septic systems	Pathogens and excess nutrients from on-site sewage disposal can pollute streams. The distance needed to separate a septic system and a river depends on the capacity of the soil to "renovate" and to handle infiltration, on how diluted the waste matter is, and on how long it takes the waste to travel from point to point.
Control stormwater runoff	Urban runoff should be pre-treated, using such means as sediment basins, catch basins, sumps, directing sheetflow over vegetation, and oil traps to minimize pollution.
Control erosion	Controlling soil erosion reduces turbidity and the amount of suspended solids. It also reduces deposition of sediment that can bury the channel bed.
Manage instream flows	The flow rates downstream of reservoirs and diversions should be adequate to meet seasonal habitat needs.
Provide fish passage	Whenever possible, dams, bridges, culverts, and channels should include means for fish to move past the structures.
Maintain water quality	Control and treat point and nonpoint sources of pollutants to minimize their volume and concentration.

TABLE 33

MAJOR WETLAND FUNCTIONS AND VALUES

Fish and wildlife values
- Fish and shellfish habitat
- Waterfowl and other bird habitat
- Furbearer and other wildlife habitat

Environmental quality values
- Water quality maintenance
  - Pollution filter
  - Sediment removal
  - Oxygen production
  - Nutrient recycling
  - Chemical and nutrient absorption
- Aquatic productivity
- Microclimate regulator

Socioeconomic values
- Flood control
- Wave-damage protection
- Erosion control
- Groundwater recharge and water supply
- Livestock grazing
- Fishing and shellfishing
- Hunting and trapping
- Recreation
- Aesthetics
- Education and scientific research

Source: "Wetlands of the United States"
United States Department of the Interior, 1984

A great blue heron searches for prey in a wetland.

Very poorly drained soils generally occur on level land or in depressions. In these areas, the water table lies at or above the surface during most of the growing season. Most of our marshes and bogs are on these soils.

Alluvial and floodplain soils occur along watercourses, occupying nearly all level areas subject to periodic flooding. Alluvial soils comprise clay, silt, sand, and gravel, or similar detrital material deposited by flowing water. Alluvial and floodplain soils range from excessively drained to very poorly drained.

Because of the ecological value of wetlands and watercourses and their sensitivity to disruption, most human activities in these areas are regulated by the state, either directly or through municipal agencies. (See appendix.) Some of these areas are also regulated by the U.S. Army Corps of Engineers.

The state Department of Environmental Protection estimates that Connecticut has lost between one-third and one-half of its original wetlands, with urban and coastal areas losing more wetland acreage than rural and upland areas. This loss, coupled with continued direct and non-direct discharges of wastewater and pollutants into numerous waterways, has had a serious impact on the ecology and water quality of many rivers and streams.

The United States Environmental Protection Agency, in its Wetlands Action Plan of 1989, called for making the preservation of wetlands, and the restoration of degraded ones, a national priority.

Wetlands Functions within River Systems

Wetlands perform many critical functions: reducing pollutants by altering them to less harmful elements or compounds, trapping sediment, recycling nutrients, controlling floods, providing wildlife habitat, recharging or discharging groundwater, and contributing to the beauty of the landscape. (See Table 33.)

Wetlands play a vital role in controlling erosion. Trees and shrubs growing along streambanks help to bind the soil, giving the banks stability. Vegetation also slows the movement of floodwater through wetland areas, reducing erosive flow velocities on floodplains.

Many wetlands, particularly those associated with rivers and streams, can temporarily detain floodwaters. The storage capacity depends on a number of factors, including topography and vegetation. Wetlands that are deep and broad release floodwaters slowly, reducing peak flow down-

FIGURE 33
HYDROLOGIC LOCATION OF WETLANDS

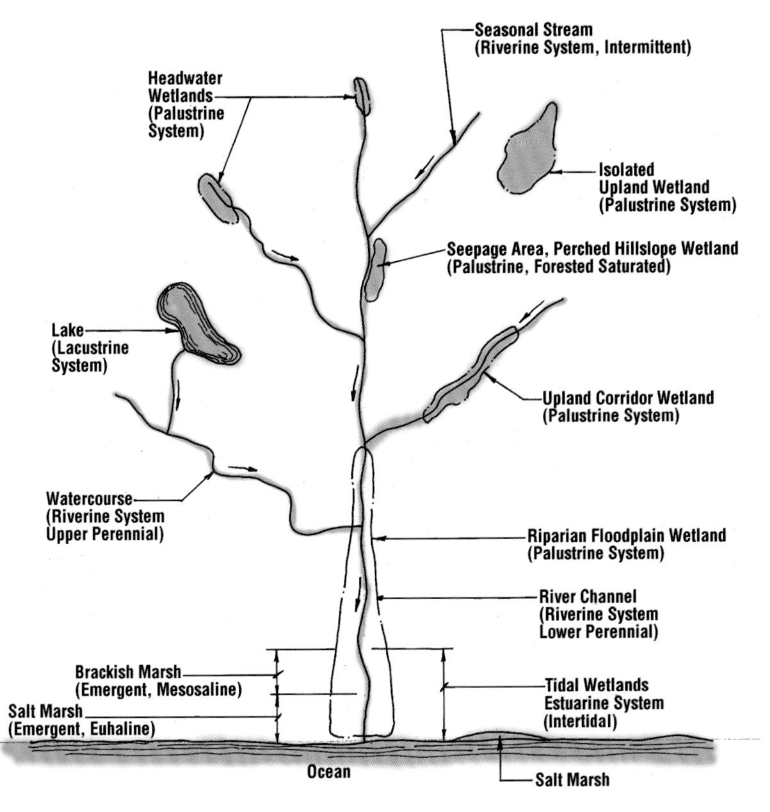

The U. S. Fish and Wildlife Service terminology is written in brackets beneath the common description.

Swift-flowing streams rich in dissolved oxygen support game fish such as trout, drawing fisherman and benefiting local economies.

Fishing is popular in urban as well as rural areas. Some people turn to fishing not as a sport, but to provide food. Unfortunately, rivers near cities handle sewage and other waste, often causing poor water quality.

stream. Such storage is especially critical in urban areas, where the potential for flood damage is high because of development.

Wetlands can improve water quality. Well-vegetated wetlands trap sediments, thereby reducing the sediment load. Many serve as "sinks" for nutrients such as nitrates and phosphates. By absorbing nutrients and, in some cases, contaminants, vegetated wetlands reduce the amount of pollutants entering surface waters.

Wetlands also provide rich habitats for fish and other wildlife. Wetland corridors not only provide homes for many species, they also

FIGURE 34

RIVERINE WETLANDS

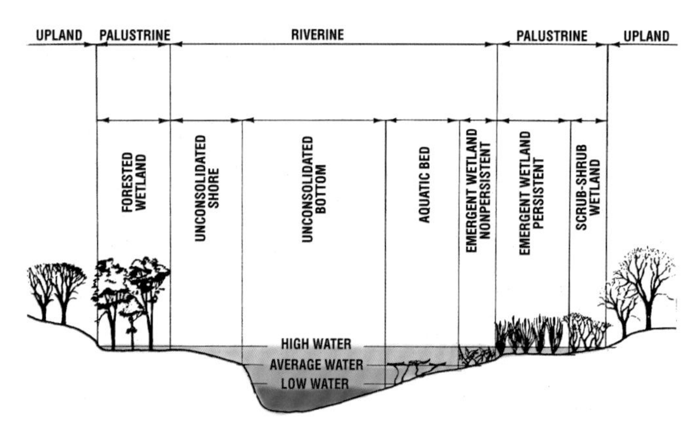

Source: "Classification of Wetlands and Deepwater Habitats of United States" U.S. Fish & Wildlife Service 1979.

Many furbearers, such as these river otters, inhabit river habitats.

serve as paths for seasonal migrations and may form the main link between larger open space areas. Some animals spend their entire life in wetland or aquatic habitats; others use them as nursery grounds or as sources of food or cover.

Among the roles played by wetlands that are more difficult to quantify is that of recharging or discharging groundwater. Many inland wetlands are discharge areas. However, those located along shores or floodplains may sometimes operate as recharge areas. (See Chapter Two.)

Topographic position has an important influence on wetland functions. The significance of principal locations is described below and in Figures 33 and 34.

Isolated Wetlands

Isolated wetlands are often found in isolated depressions or on hillsides with groundwater discharges. Although not directly connected with stream systems, they may retain some surface runoff and reduce the peak flow of streams and rivers lower in the watershed. Upland wetlands also provide landscape diversity and support wildlife.

A unique type of upland wetland is the vernal pool. Vernal pools are temporary water bodies enclosed within small depressions without permanent outlets. They are wet primarily in the spring, particularly during and after snowmelt. Since they dry up during the summer, they cannot support fish or most other predators. Because of this, they are home to specialized invertebrates, such as fairy shrimp, and breeding habitat for certain salamanders, frogs and toads, many of which are dependent on these temporary wetlands for their survival.

Headwater Wetlands

Frequently, in the upper parts of a watershed, there are groundwater discharge wetlands that flow directly into the headwater streams that eventually become the main channels of the watershed. Such wetlands are often on slopes with an impervious subsoil that forces groundwater to flow to the surface. This type of wetland often is important in maintaining streamflow during dry periods, when it discharges groundwater into the channel; it has little, if any, groundwater-recharge capability.

Alluvial Wetlands (Floodplains)

Alluvial wetlands are composed of sedimentary floodplain soils with a wide variety of drainage classes. In Connecticut, all alluvial soils are regulated as wetlands, whereas under federal definitions only poorly and very poorly drained soils are considered wetlands.

Floodplains are capable of storing large quantities of floodwater and conveying it more gradually, thereby helping to reduce the peak flows in the main channel and farther downstream.

At times of normal or low streamflow, floodplains serve as points where groundwater is recharged when direct rainfall infiltrates the permeable soils. Such regions often are important wildlife habitats. They are also used extensively for agriculture and recreation.

TABLE 34
COMPARISON OF WETLAND REGULATIONS

CONNECTICUT REGULATIONS	
Regulating authority	Individual municipalities
Related agencies	Connecticut Department of Environmental Protection, Inland Water Resources Management Division (if requested by local agency)
Definition of a wetland	***Soil types*** that are poorly drained, very poorly drained, alluvial, and floodplain. ***Watercourses*** including: rivers, streams, brooks, marshes, swamps, bogs, and all other bodies of water, natural or artificial, public or private, which are contained within, flow through, or border the state of Connecticut.
Areas that are regulated	Soil boundary of wetlands. Limits of watercourse. A buffer zone around wetlands as and if established by local regulations.

FEDERAL REGULATIONS	
Regulating authority	United States Army Corps of Engineers
Related agencies	U.S. Environmental Protection Agency, U.S. Fish and Wildlife Service
Definition of a wetland	***Hydrophytic vegetation*:** Land supports predominantly hydrophytes. ***Hydric Soils:*** Those meeting criteria set by the National Technical Committee for Hydric Soils. ***Hydrology:*** Saturation to the surface or inundated at some time during an average rainfall year.

Area described in the "Federal Manual for Identifying and Delineating Jurisdictional Wetlands." January, 1989; and Corps of Engineers 1987 manual

WETLAND HABITATS

There are several types of wetland habitats, each with a specific combination of soil, water, vegetation, and wildlife. The most important factors influencing freshwater wetlands are how much water is present, for how long, when during the year, and — when there is no standing water — the degree of soil moisture. These hydrological factors affect the mix of vegetation and wildlife. Wetlands can be classified by their predominant vegetation as wet meadows, marshes, swamps, fens, or bogs.

Wet Meadows

Wet meadows are wetlands with poorly drained soils that are covered with grasses, sedges, rushes, and other emergent vegetation. They may have shallow surface water or runoff on them during winter and spring. Most wet meadows are former or active agricultural land that has been cleared of trees. They often occur on floodplains.

Marshes

Marshes are a type of wetland where the ground is submerged during the growing season by six inches to three feet of standing water with little flow velocity. The water limits vegetation to soft-stemmed, emergent species. Typical plants include cattail, reeds, arrowhead, pickerel weed, and wild rice. Deeper marshes may have areas of open water with rooted, floating plants such as water lily. "Deadwood swamps" are usually new marshes where the water depth has increased enough to kill the trees relatively recently. A valuable habitat for waterfowl such as ducks, geese, and herons, marshes are also used by many amphibians and some mammals.

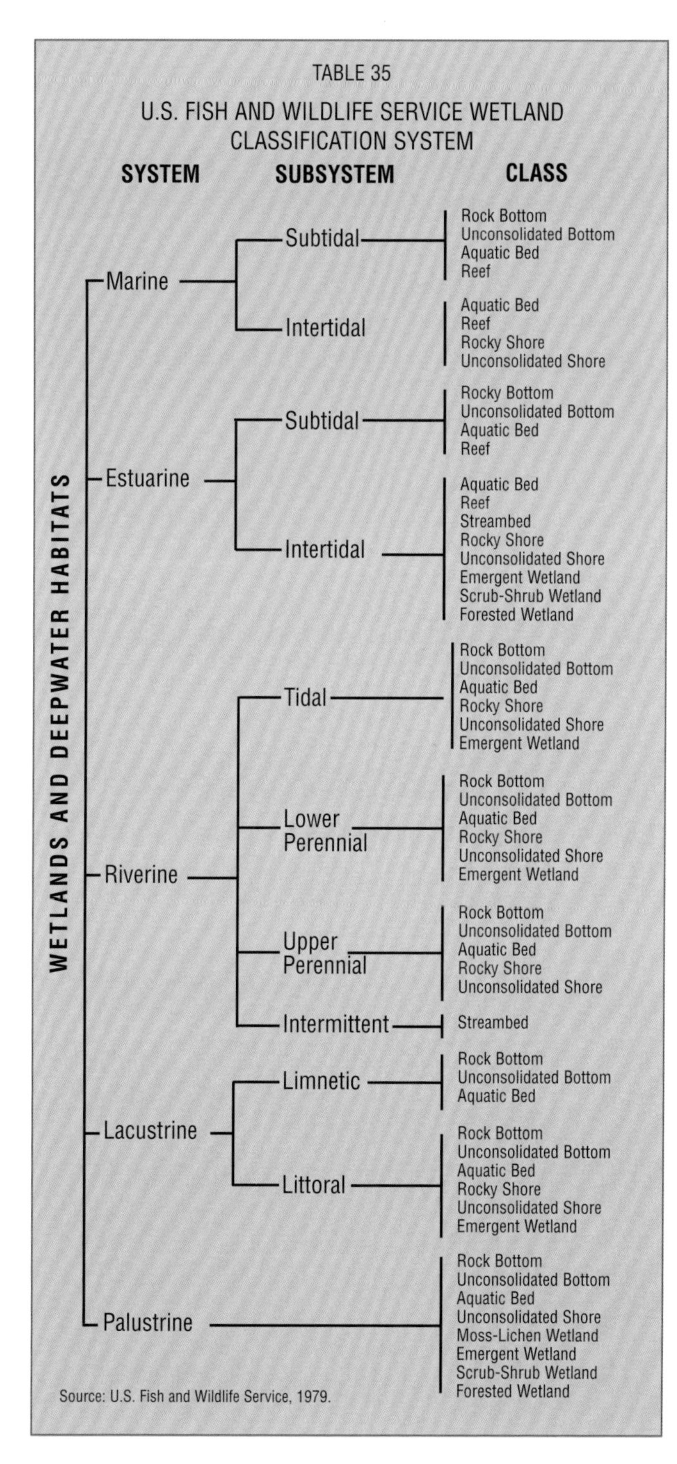

TABLE 35

U.S. FISH AND WILDLIFE SERVICE WETLAND CLASSIFICATION SYSTEM

WETLANDS AND DEEPWATER HABITATS

SYSTEM	SUBSYSTEM	CLASS
Marine	Subtidal	Rock Bottom Unconsolidated Bottom Aquatic Bed Reef
	Intertidal	Aquatic Bed Reef Rocky Shore Unconsolidated Shore
Estuarine	Subtidal	Rocky Bottom Unconsolidated Bottom Aquatic Bed Reef
	Intertidal	Aquatic Bed Reef Streambed Rocky Shore Unconsolidated Shore Emergent Wetland Scrub-Shrub Wetland Forested Wetland
Riverine	Tidal	Rock Bottom Unconsolidated Bottom Aquatic Bed Rocky Shore Unconsolidated Shore Emergent Wetland
	Lower Perennial	Rock Bottom Unconsolidated Bottom Aquatic Bed Rocky Shore Unconsolidated Shore Emergent Wetland
	Upper Perennial	Rock Bottom Unconsolidated Bottom Aquatic Bed Rocky Shore Unconsolidated Shore
	Intermittent	Streambed
Lacustrine	Limnetic	Rock Bottom Unconsolidated Bottom Aquatic Bed
	Littoral	Rock Bottom Unconsolidated Bottom Aquatic Bed Rocky Shore Unconsolidated Shore Emergent Wetland
Palustrine		Rock Bottom Unconsolidated Bottom Aquatic Bed Unconsolidated Shore Moss-Lichen Wetland Emergent Wetland Scrub-Shrub Wetland Forested Wetland

Source: U.S. Fish and Wildlife Service, 1979.

Swamps

Swamps are wetlands dominated by shrubs and trees. They can have as much as a foot of standing water on a seasonal basis and high groundwater most of the year. Swamps are often located alongside watercourses and on floodplains. Some wooded swamps are located in headwater areas and are supported by groundwater discharge with little surface water. Red maple is almost always the most common tree, with lesser numbers of American elm, ash, and pin oak. On floodplains, silver maple can often be the dominant tree. Common shrubs include spicebush, highbush blueberry, speckled alder, and briars. Swamps are used extensively by upland mammals and birds.

Bogs and Fens

Bogs and fens are peatlands with a well-developed moss carpet, usually dominated by sphagnum mosses. In Connecticut, they can be found in glacial kettles or other depressions within deposits of sands and gravels, in bedrock depressions, or on the margin of nutrient-poor lakes, ponds or slow-flowing streams. These peatlands form where the rate of plant-material accumulation exceeds the rate of decay. Under these conditions, depressions fill with water-saturated, slowly decaying organic plant remains called peat, rather than inorganic silts and clays.

Bogs generally have a dominant cover of dwarf shrubs such as leatherleaf and stunted trees such as black spruce. Fens, in contrast, are dominated by carpets of sedges and other grass-like plants intermixed among the mosses. A particularly uncommon type of fen occurs in the marble valleys of western Connecticut on spring-fed slopes where the nutrient-poor, alkaline waters provide habitat for a large number of endangered and threatened species.

RIPARIAN ZONES

The riparian zone is the area adjacent to a river, a transition between the stream and its upland. It may consist of wetlands, relatively level upland, or steep hillsides that slope to the water's edge. It may be developed or undeveloped land.

Even if riparian areas are relatively dry and are thus not strictly wetlands, they are critical to the entire river. Riparian vegetation is the main source of organic detritus for headwater streams, and is thus the basis of the food chain. This zone also helps shade the water and provide cover for both fish and terrestrial animals.

The amount and type of vegetation is influenced by the watercourse. At river's edge, plant life is subject to flooding, which causes both erosion and deposition of sediment. The damage to riverbank vegetation by winter ice can also be an influence. For example, black willow, a species that grows rapidly, is often found along stream sides. If uprooted or broken by floating debris, it can send forth new roots to gain a new foothold. In this manner, the trees help to stabilize riverbanks.

Sometimes a mix of species will occur with great regularity along the entire length of a particular corridor. For example, one stream system may have sandy shores dominated by annual herbs and grasses. Another may have a dense variety of shrubs, including speckled alder, pussywillow and red-osier dogwood. Others might be lined with box elder, black willow, and white ash.

Riparian streambelts provide a continuous "edge" habitat, the delineation between two distinct habitats — in this case terrestrial and aquatic. The mix of wildlife species is especially diverse in edge habitats.

In many instances, the corridor formed by a stream and its riparian zone provides a continuous habitat that serves the needs of many species. In addition to drinking water, streambelts can offer protected sites for nests or dens; food sources, especially if fruit-producing plants are present; and a corridor for safe travel. Such corridors also provide important links between larger habitat areas such as wildlife reserves.

Clearing, grading, impervious ground cover, installing septic systems, and over-pumping nearby wells all can significantly damage riparian zones.

Preserving a riparian buffer of sufficient width is critical to maintaining the ecological health of a river.

Variables to consider in determining the optimum buffer width include soil types, land slope, density of vegetation, and upland land-use activities.

A moderately wide buffer generally is enough to shade the water, provide organic detritus and provide a root mass to stabilize the banks. A wider buffer provides more filtering of storm runoff, increases floodwater storage and conveyance capacity, and provides more wildlife habitat. Typically, buffers range between 50 and 100 feet.

TABLE 36

RIPARIAN ZONE FUNCTIONS

- Shades water, reduces temperature
- Filters upland runoff
- Stabilizes channel banks through root massing
- Provides shelter and cover for wildlife
- Provides shelter for fish
- Contributes organic detritus to channel
- Conveys flood flows
- Provides recreation and aesthetic variety
- Recharges groundwater
- Serves as corridor for animal movement

FIGURE 35

CROSS SECTION OF RIPARIAN ZONE

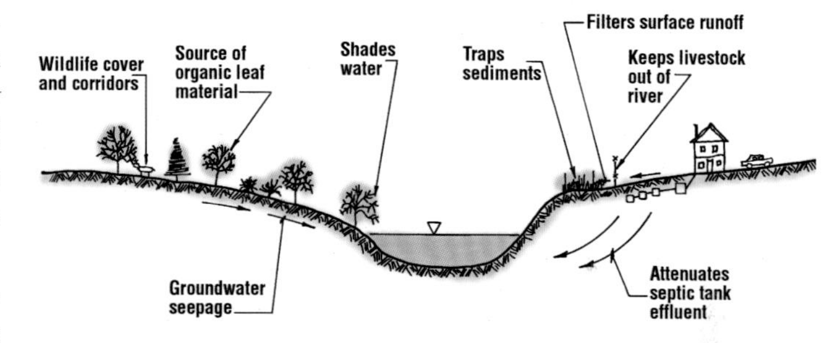

Muskrats dwell exclusively in wetland areas.

BIRDS

Tables derived from: DeGraaf, Richard M., and Rudis, Deborah D. *New England Wildlife: Habitat, Natural History and Distribution.* USDA Forest Service, Northeastern Forest Experiment Station. Gen.-Tech. Report NE-108

Name	What it needs for a habitat	What it eats	Notes of interest
PIED BILLED GREBE	Open water in sluggish streams with well-vegetated banks	Fish, crustaceans	Often observed swimming, diving
AMERICAN BITTERN	Undisturbed area with tall marsh vegetation	Frogs, reptiles	Observe by standing and waiting; walk slowly
GREAT BLUE HERON	Shores of streams; wooded swamps	Insects, fish, amphibians, crustaceans	Observe by standing and waiting; walk slowly
GREEN-BACKED HERON	Nearly all freshwater and saltwater habitats including rivers and streams	Small fish, crustaceans, mollusks, insects	Observe by standing and waiting; walk slowly
CANADA GOOSE	Shallow water with abundant plant foods, winters on ice-free rivers	Shoots of grasses, sedges and marsh plants	Feeds by grazing, immersing head & neck
WOOD DUCK	Shallow waters along streams; marshes; trees with large cavities for nesting	Insects, acorns, seeds of aquatic plants	Feeds by grazing, immersing head & neck
AMERICAN BLACK DUCK	Marshy borders of ponds, lakes, rivers, and streams	Mollusks, submerged aquatic plants, seeds of marsh plants, invertebrates	Often observed dabbling, grazing, gleaning
MALLARD	Ponds, lakes, rivers, wet meadows, marshes, wooded swamps	Seeds of sedges, grasses and smartweeds	Feeds by dabbling, grazing, gleaning
COMMON MERGANSER	Winters in fresh and brackish waters of rivers, lakes and ponds	Fish, mollusks	Feeds by diving, immersing head

BIRDS

Name	What it needs for a habitat	What it eats	Notes of interest
OSPREY	Near large bodies of water with abundant fish; coast, rivers, lakes	Fish	Feeds by hovering and diving into water
BALD EAGLE	Winters along coast, ice-free rivers with abundant fish	Fish, small mammals, birds, turtles, carrion	Frequently observed soaring, hawking, hovering, and diving
VIRGINIA RAIL	Freshwater marshes with sedge and cattail edge	Small fish, insects, seeds of marsh plants, invertebrates	Feeds by probing, gleaning
SPOTTED SANDPIPER	Freshwater along edges of ponds, lakes, rivers; prefers open terrain	Insects, small fish and crustaceans	Often observed gleaning, swimming, diving, and tail bobbing.
AMERICAN WOODCOCK	Moist woodlands, bottomlands, swamps, streambanks	Earthworms, beetle larvae	Feeds by probing, gleaning
BARRED OWL	Low wet, deep woods, wooded swamps	Mice, small mammals, frogs, crayfish	Feeds by swooping and pouncing
BELTED KINGFISHER	Banks near rivers and streams for nest sites	Fish, crayfish and crustaceans	Feeds by diving, skimming water surface
EASTERN PHOEBE	Cliffs, ledges near streams	Flying insects	Feeds by seeking flying insects
TREE SWALLOW	Farmlands, river bottomlands, ponds, wooded swamps or marshes with dead standing trees	Flying insects	Feeds by hawking, skimming water surface
NORTHERN ROUGHWINGED SWALLOW	Any open area with adequate nest sites and nearby water supply; river valleys	Flying insects	Feeds by hawking, skimming water surface

BIRDS

Name	What it needs for a habitat	What it eats	Notes of interest
BANK SWALLOW	Riverbanks, gravel pits, sand or clay banks near water	Flying insects	Feeds by hawking, skimming water surface
MARSH WREN	Large, fresh or brackish marshes with tall emergent vegetation; shores of sluggish rivers	Insects, spiders	Feeds by gleaning, hawking
NORTHERN WATERTHRUSH	Wooded swamps, bogs, woodland streams	Aquatic insects, beetle larvae, mosquitoes	Feeds by ground gleaning
LOUISIANA WATERTHRUSH	Bottomland forests; wooded valleys of rocky brooks and fast-flowing streams	Dragonfly and cranefly larvae; beetles, ants, caterpillars	Feeds along ground surface
COMMON YELLOWTHROAT	Swampy thickets on margins of damp woods and woodland streams	Cankerworms, fall webworms, grasshoppers and other insects	Observed in shrubs and ground gleaning
SWAMP SPARROW	Marshes, swamps, bogs, low swampy shores of streams and rivers	Insects, seeds	Observed wading, feeds by gleaning
RED-WINGED BLACKBIRD	Marshes, swamps, wet meadows with extensive tall emergent vegetation	Insects, weed seeds, grain	Feeds by ground gleaning

MAMMALS

Name	What it needs for a habitat	What it eats	Notes of interest
WATER SHREW	Marshes, shrub zones along ponds and streams in coniferous forests	Larvae of aquatic insects	Uncommon; secretive behavior
BEAVER	Brooks, streams, rivers often bordered by woodland	Bark of deciduous trees, particularly aspen, alder, willow, birch, and maple	Forms lifelong monogamous pair bond; constructs dams to retain water, and lodges for shelter and rearing young
MUSKRAT	Marshes, especially cattail; shallow lakes and ponds; sluggish streams	Aquatic plants, including reeds, pondweeds and bulrushes; freshwater clams	Dens in streambanks or may construct dome-shaped lodge of robust emergent vegetation
RACCOON	Wooded areas near fields and watercourses	Omnivorous; animal matter including crayfish, worms and carrion; fruits, seeds and tender shoots	Primarily nocturnal; alert, intelligent animals; usually dens in trees 10 feet or more above ground
ERMINE	Woods and thickets with dense brushy cover, often near watercourses	Mice, chipmunks, moles, or shrews	Nocturnal; molts to white in winter; usually dens below ground under fallen tree or stump
LONG TAILED WEASEL	Open woods, grassland, river bottomlands	Small mammals, including voles, mice, rabbits, and shrews; some ground-nesting birds	Active yearround; sometimes turns white in winter
MINK	Streambanks, lakeshores and marshes; favors forested wetlands with dense cover	Small mammals including rabbits, moles, muskrats; also fish, frogs, salamanders, crayfish, and clams	Mainly nocturnal; active yearround; dens in hollow logs, under tree roots or burrows along streambanks
RIVER OTTER	Borders of streams, lakes, and other wetlands in forested areas.	Aquatic animals including fish, frogs, crayfish, salamanders, and turtles; also birds, small mammals and invertebrates	Most active in early morning and evening; may use abandoned beaver or muskrat lodge for den site
WHITE-TAILED DEER	Forest edges, swamp borders, forested river corridors	Woody, deciduous plants; some coniferous growth	Gregarious, forms small groups; requires dense cover for winter shelter and adequate browse

AMPHIBIANS

Name	What it needs for a habitat	What it eats	Notes of interest
NORTHERN DUSKY SALAMANDER	Woodlands at edge of clear, running water; favors clear, rocky streams	Small aquatic invertebrates including grubs, worms, and crustaceans	Larvae stage riparian; eggs laid June-September and guarded by female; under logs and in moss near water
NORTHERN TWO-LINED SALAMANDER	Floodplain bottoms, moist forest floors; along brooks and streams, boggy areas near springs and seeps	Insects, particularly beetles, beetle larvae, mayflies, stonefly nymphs, and dipterans	Travels during wet nights; eggs deposited in clusters attached to bottoms of logs or stones in running water
BULLFROG	Shorelines of large bodies of water with emergent vegetation, lakes, river oxbows	Fish, other frogs, newts, salamanders, young turtles, mice, crayfish	Bullfrogs are very susceptible to pollution; remain as tadpoles for 2-3 winters
GREEN FROG	Riparian, inhabiting margins of streams, lakes, creeks, ponds, vernal pools, and moist woodland; seldom far from water	Insects and larvae, worms, small fish, crustaceans, and mollusks	Eggs laid in floating masses of jelly attached to underwater branches; tadpoles generally overwinter for 1-2 years
PICKEREL FROG	Clear streams and shores of lakes and ponds	Terrestrial arthropods; also snails, small crayfish	Sensitive to pollution and other changes in water quality

Bullfrogs require larger bodies of water for their habitat.

REPTILES

Name	What it needs for a habitat	What it eats	Notes of interest
SPOTTED TURTLE	Unpolluted shallow bodies of water including woodland streams, ponds, marshes	Omnivorous; eats mollusks, crustaceans, spiders, earthworms, and aquatic insects	Uncommon to rare in New England; loss of habitat and over-collecting may have caused decline
WOOD TURTLE	Wooded riverbanks; frequents slow, meandering streams with sandy bottoms and overhanging alders	Omnivorous; eats young plants, mushrooms, berries, insects, worms, slugs, tadpoles, frogs, and fish	Once common, population declining throughout its range
EASTERN PAINTED TURTLE	Quiet ponds, marshes, woodland pools, rivers, lake shores, and slow-moving streams	Aquatic insects, snails, small fish, mussels, carrion, and aquatic plants	Basks on logs or rocks, often in large groups
NORTHERN WATERSNAKE	Aquatic habitats; common near bridges and spillways where rocks provide cover	Fish, frogs, toads, salamanders, insects, and crayfish	Frequently found basking; active day and night

Wood turtles prefer to live on the wooded banks of slow-flowing streams.

FISH

Name	What it needs for a habitat	What it eats	Notes of interest
BROOK TROUT	Small to medium-size upland or lowland streams having high water quality and cold temperatures; adjacent wetlands and groundwater; abundant instream and riparian cover; clean gravel for spawning	Aquatic and terrestrial invertebrates (primarily aquatic insects); small fish	Highly sought-after gamefish; considered by many to be the most beautifully colored of all New England fishes
BROWN TROUT	Small to large upland or lowland streams having high water quality and cold temperatures; adjacent wetlands and groundwater; abundant instream and riparian cover; clean gravel for spawning	Aquatic and terrestrial invertebrates (primarily aquatic insects); fish	Highly sought-after gamefish; the preferred target of the dry-fly enthusiast
WHITE SUCKER	Small creeks to large rivers; often makes spawning runs into streams from lakes and larger rivers	Zooplankton, invertebrates and small amounts of plant material	Most abundant fish species by weight in Connecticut streams
FALLFISH	Small to moderate-size streams; clean gravel for constructing nests; instream cover	Plankton, aquatic insects and small amounts of plant material; typically feeds on drift in mid-water column	Statewide distribution; nests are built in spring and consist of piles of gravel up to 4 feet long and 12 to 24 inches high
CREEK CHUB	Small to moderate-size streams; clean gravel for constructing nests; instream cover	Plankton, aquatic insects and small amounts of plant material	Similar to fallfish in appearance and life history; populations mostly restricted to western Connecticut
COMMON SHINER	Small to moderate-size streams; instream cover	Plankton, aquatic insects and small amounts of plant material	Widespread distribution; spawns over nests of other species, including fallfish and creek chubs

FISH

Name	What it needs for a habitat	What it eats	Notes of interest
SPOTTAIL SHINER	Large rivers or reservoirs	Plankton, small invertebrates and some plant material	Most abundant fish species in large rivers (e.g., Connecticut River); important forage species
CUTLIPS MINNOW	Small to moderate-size streams with boulder substrate and little vegetation	Bottom-dwelling aquatic insects and mollusks; small amounts of plant material	Highly specialized trilobed lower lip; found principally in western Connecticut
GOLDEN SHINER	Slow-moving water in streams and larger rivers; abundant vegetation	Midwater zooplankton, insects and filamentous algae	Spawns over vegetation; important forage species for large gamefish
BRIDLE SHINER	Warm, slow-flowing waters	Small bottom dwelling invertebrates	Rare in Connecticut
BLACKNOSE DACE	Small to moderate-size clear streams; riffles	Small aquatic insects	Widespread and abundant; often only fish in very small waters; frequently inhabits headwater streams with brook trout
LONGNOSE DACE	Riffles in small to moderate-size, clear streams	Small aquatic insects; bottom feeder	Widespread
CARP	Larger streams and rivers; slow-moving warm water	Benthic invertebrates and algae	Largest minnow found in Connecticut
CATFISH (FOUR SPECIES)	Slow-moving, large streams and rivers; bullheads in small to medium streams	Invertebrates and fish	Abundant in large rivers
REDBREAST SUNFISH	Slow-moving streams and rivers	Invertebrates and small fish	Most common riverine sunfish. Other sunfish species such as bluegill, rock bass and black crappie inhabit larger rivers, streams and smaller waters downstream of impoundments

FISH

Name	What it needs for a habitat	What it eats	Notes of interest
RAINBOW TROUT	Small to large sized upland or lowland streams having high water quality and cool temperatures; adjacent wetlands and groundwater; abundant instream and riparian cover; clean gravel for spawning	Invertebrates (primarily aquatic insects)	Most Connecticut rainbows are stocked fish; only a few wild populations have been identified
ATLANTIC SALMON	Small to large upland or lowland streams having high water quality and cool temperatures, gravel/boulder substrate, direct downstream access to sea	Invertebrates (primarily aquatic insects)	Extinct in Connecticut by 1800; restoration program under way
AMERICAN EEL	Large rivers to small streams; access to sea	Invertebrates and small fish	Highest densities in streams adjacent to Long Island Sound or just upstream from large rivers; spends much of time buried in gravel or rock substrate
SLIMY SCULPIN	Cold, clear gravel-bottomed streams	Aquatic insects	Often in association with brook trout; 5 inches long
TESSELLATED DARTER	Small to large streams; sand bottom	Small invertebrates	Statewide distribution; 2-3.5 inches long
SWAMP DARTER	Slow-moving streams; aquatic vegetation	Small invertebrates	Rare; 2 inches
SEA LAMPREY	Soft-bottom areas in small to medium-size streams; direct downstream access to salt water	Plankton in fresh water; parasitic on larger fish in salt water	Anadromous: Spawn in freshwater, migrate to sea after 5+ years

FISH

Name	What it needs for a habitat	What it eats	Notes of interest
AMERICAN BROOK LAMPREY	Sandy or mud bottom, slow-water areas in small to medium-size streams	Plankton and detritus	Entire life cycle in fresh water
RIVER HERRING	Large rivers; alewives and blue-back herring will spawn in small to medium-size tributaries; open access to the sea	Zooplankton	Alewife, blueback herring and American shad are anadromous; gizzard shad spend entire life cycle in fresh water; life history of hickory shad is largely unknown
FOURSPINE STICKLEBACK	Coastal streams and tributaries to large rivers; abundant aquatic vegetation	Plankton	Constructs nests of plant material
BANDED SUNFISH	Slow, tannic waters with low pH	Small invertebrates	Rare; distribution limited to eastern Connecticut; 1-3 inches long
SMALLMOUTH BASS	Moderate current in larger streams and rivers; rocky substrate	Insects, crayfish and fish	Popular gamefish
REDFIN PICKEREL	Weedy areas in slow-moving streams	Insects, crayfish and fish	A related species, chain pickerel, frequents waters downstream from impoundments
LARGEMOUTH BASS	Large rivers	Insects, crayfish and fish	Popular gamefish; occasionally in smaller waters downstream from impoundments
WHITE PERCH	Large rivers, including estuarine areas	Zooplankton, invertebrates and fish	Can tolerate wide range of salinities

Surface Water Quality and Its Management

Accuse not Nature, she hath done her part; do thou but thine.

John Milton, *Paradise Lost*

Flotation booms used to absorb oil.

Metropolitan District filtration plant

Surface Water Quality and Its Management

Water is one of the most important, and visible, gauges of overall environmental quality.

About 80 percent of Connecticut's drinking water is obtained from surface sources. Humans also need high-quality water for recreational uses, irrigation, and industry. Wildlife, especially fish and other aquatic species, also require water of suitable quality.

Water in streams and rivers can affect other resources. For example, it affects the quality of groundwater where surface waters recharge aquifers, and has a direct impact on the health of riparian wetland environments. Ultimately, since all Connecticut rivers eventually discharge into coastal regions, they influence the water quality in Long Island Sound.

The water quality in Connecticut's rivers is believed to have declined rapidly with the increased industrialization and growth of cities after 1850. Sewer systems were installed in urban areas, discharging both human wastes and surface runoff directly into rivers; factories and mills used waterways to dispose of untreated machine oil, solvents, dyes, and other chemicals. Rivers also were used to dispose of garbage.

A State Sewer Study Commission, appointed in 1897, recommended basic primary treatment of municipal sewage. However, little was done and many rivers were reported to be grossly polluted.

Connecticut's Clean Water Act of 1967 recognized the serious problem and mandated a plan to improve water quality. Municipalities and industries were ordered to provide wastewater treatment and to eliminate overflows from combined systems that carried both wastewater and stormwater.

Discarded tires, shopping carts, and other debris litter many urban rivers.

TABLE 37

CONNECTICUT SURFACE WATER CLASSIFICATIONS (1997)

CLASS	DESIGNATED USE	COMPATIBLE DISCHARGES
AA	Existing or proposed public drinking water supply impoundments and tributary surface waters	Treated backwash from drinking water treatment facilities; minor cooling water; clean water
A or SA*	May be suitable for drinking water supply (Class A), may be suitable for all other water uses including swimming, shellfish resource; character uniformly excellent; may be subject to absolute restrictions on the discharge of pollutants	Treated backwash from drinking water treatment facilities; minor cooling water; clean water
B or SB*	Suitable for swimming, other recreational purposes, agricultural uses, certain industrial processes, and cooling; excellent fish and wildlife habitat; good aesthetic value	Those allowed in Class AA, A; major and minor discharges from municipal and industrial wastewater treatment
C or SC*	May have limited suitability for certain fish and wildlife, recreational boating, certain industrial processes, and cooling; good aesthetic value; not suitable for swimming. Quality considered unacceptable; goal is B or SB	Same as B or SB
D or SD*	May be suitable for swimming or other recreational purposes, certain fish and wildlife habitat, certain industrial processes, and cooling; may have good aesthetic value. Present conditions, however, severely inhibit or preclude one or more of the above resource values. Quality considered unacceptable; goal is B or SB	Same as B or SB

* Designates salt or brackish water.

The Clean Water Act, first adopted in 1948 but significantly amended in 1972 (Federal Water Pollution Control Act of 1972) and periodically revised, reinforces and supports Connecticut's actions and provides some federal funding for programs to improve water quality. The law's objective is to restore and maintain the chemical, physical, and biological integrity of the nation's waters, and it regulates both municipal and industrial discharges.

The 1987 amendment to the Clean Water Act requires states to develop management programs for nonpoint sources of pollution, supplementing the traditional emphasis on point sources, and requires water quality standards for more than 100 chemicals.

The Federal Safe Drinking Water Act and Surface Water Treatment Rule also affect the management of watersheds and rivers used for drinking water supplies, requiring all public water systems using surface water to remove or kill disease-causing microorganisms. These regulations emphasize improved watershed protection and water filtration.

Today, the Connecticut Departments of Environmental Protection and Public Health monitor the state's water resources and work together to protect and improve water quality.

WATER QUALITY STANDARDS

Operating under statutes that require it to protect the public health and welfare and promote the economic development of the state, the Department of Environmental Protection has adopted water-quality standards for both surface water and groundwater. The policy is to restore and maintain waters so as to protect public health, to provide safe water for drinking and other uses, to provide wastewater assimilation, and to foster recreation. The standards, which were first adopted in 1970 and are periodically updated, provide guidance for allowing new wastewater discharges and assigning priorities for funding cleanup efforts and other projects.

Environmental officials have categorized the state's surface waters into five classifications, ranging from AA, the cleanest, through Class D, those with the poorest water quality. (See Table 37.) The Connecticut goal is to get all waters to at least B or SB quality.

A state anti-degradation policy requires the Department of Environmental Protection to maintain existing high-quality waters at their current levels of cleanliness and prohibits any lowering of water quality that would impair present uses.

In the past three decades, there have been significant improvements in the quality of Connecticut's surface water, thanks largely to the upgrading of wastewater treatment plants, construction of new plants, and greater control of discharges by industry.

The 1996 Department of Environmental Protection Water Quality Report to Congress says that 65 percent of the 893 miles of major rivers monitored here met the current goals. An additional 27 percent partially met them. Eight percent did not meet the standards.

The goal that a stream be clean enough for fishing has been met in 668 miles of main rivers, and the "swimmable" goal has been achieved in 645 miles.

The success of Connecticut's water quality program is highlighted by the Naugatuck and Willimantic Rivers. Heavily polluted just 20 years ago, both now exhibit acceptable aesthetic quality and support increasing populations of fish.

WATER QUALITY CHARACTERISTICS

The quality of water is measured in terms of its physical, chemical, and biological characteristics. It is affected both by natural conditions and human activities. (See Table 38.)

The physical characteristics — such as color, odor, and turbidity — are readily apparent and form our first impression of water quality. Chemical characteristics, while seldom visible, often affect the water's toxicity and ability to support aquatic life. Biological characteristics are critical because some organisms can cause disease, while others can affect the water's taste and odor.

Water is seldom absolutely pure, but usually contains various suspended and dissolved materials. Natural water quality varies in response to rainfall patterns and atmospheric pollutants, as well as the type of soils, rocks and vegetation in the watershed. Natural contaminants can be in the form of minerals dissolved by rainfall that percolates through the soil. These can include salt, calcium, iron, manganese, and aluminum. Some rocks, such as limestone and sandstone, are easily dissolved or eroded and add to the concentration of chemicals in a stream or its load of suspended sediments.

Dissolved oxygen is critical for supporting aquatic life. Surface waters receive their oxygen directly from the atmosphere, from photosynthesis of aquatic plants, or from relatively oxygen-rich surface

TABLE 38

SELECTED WATER QUALITY INDICATORS

PHYSICAL	CHEMICAL	BIOLOGICAL
Color	Oxygen	Indicator bacteria
Odor	Metals	Algae
Temperature	Nitrogen	Benthic macroinvertebrates
Turbidity	Phosphorus	Fish
Suspended solids	pH level	
Dissolved solids	Conductivity	
Volatile solids	Hydrocarbons	
	Pesticides	
	Biochemical oxygen demand	

TABLE 39

SOURCE SUMMARY REPORT – RIVERS AND STREAMS
(880 TOTAL STREAM MILES ASSESSED)

SOURCE CATEGORIES	DEGREE OF IMPACT AND NUMBER OF RIVERS AFFECTED					
	MAJOR	#	MOD./MINOR	#	THREATENED	#
POINT SOURCE:						
Industrial	69.0	3	48.0	4	10.0	1
Municipal	212.0	9	8.0	1	30.0	4
Municipal Pre-Treatment	0.0	0	44.0	1	0.0	0
Combined Sewer Overflow	126.0	10	11.0	1	0.0	0
NONPOINT SOURCE:						
Unspecified NPS	0.0	0	129.0	3	0.0	0
Agriculture	0.0	0	102.0	4	92.0	10
Highway, Bridge Const.	0.0	0	16.0	1	0.0	0
Land Development	0.0	0	0.0	0	19.0	3
Urban Run-off	0.0	0	97.0	11	22.0	2
Storm Sewers	0.0	0	51.0	4	7.0	2
Dredge Mining	0.0	0	44.0	1	0.0	0
Landfills	0.0	0	108.0	6	99.0	13
On-site Wastewater System	13.0	2	3.0	1	56.0	5
Channelization	10.0	2	44.0	1	0.0	0
Dam Construction	0.0	0	25.0	1	0.0	0
Removal of Riparian Vegetation	0.0	0	44.0	1	20.0	3
Highway Maintenance, Runoff	0.0	0	0.0	0	20.0	3
In-Place Contaminants	68.0	2	0.0	0	16.0	2

Reproduced from Connecticut DEP Report "Nonpoint Source Pollution, an Assessment and Management Plan" Feb. 28, 1989

TABLE 40

RELATIVE TURBIDITY RANKINGS

LOW	Stream bottom is distinctly visible in four feet of water.
MODERATE	Stream bottom is indistinct in four feet of water.
HIGH	Stream bottom is visible only where water is less than one foot deep.

runoff and tributaries, where the mixing of air and water by turbulence causes a greater rate of reaeration than in sluggish and stagnant water.

Much of the oxygen produced by photosynthesis comes from algae living in the water. The oxygen concentration in waters with heavy algae blooms often follows the photosynthetic cycle, peaking in the afternoon and declining in the evening. However, such blooms are eventually followed by large die-offs of algae, and the decaying organic material causes excessive consumption of oxygen, which can lead to fish kills. The amount of oxygen utilized during the decomposition of organic material is called the biochemical oxygen demand (BOD).

Algae blooms, and the decreased water clarity and depleted oxygen levels that result, are stimulated by excessive levels of nutrients, such as phosphorus and nitrogen, which may have either natural or human sources. The latter include septic systems, sewage treatment plants, farm and lawn fertilizer, and animal wastes.

Water temperature affects aquatic habitat even in the absence of pollution. Fish and other organisms are sensitive to temperature, generally preferring a specific range. Also, cooler water retains more oxygen than warmer water does. The temperature range of a small stream can rise if trees are removed from its banks, increasing the amount of sunlight it receives.

Turbidity refers to the water's clarity, the ability of light to penetrate. Water with high turbidity is not clear, and is less attractive. Turbidity levels increase as a result of soil erosion, urban runoff and algae. The suspended and dissolved particles that cause turbidity often absorb chemicals or support bacteria. High turbidity inhibits aquatic life, can cause unpleasant color and odor, and makes a river less desirable for recreation.

The pH level of water, its acidity or alkalinity, is affected by natural factors, such as the acidity of soils and rock, and by human influences, such as wastewater discharges and airborne deposits from the burning of fossil fuels, which generates sulfur and nitrogen compounds that contribute to acid rain. High and low pH values limit aquatic life and cause corrosion.

Natural biological pollutants include living organisms and their decomposing elements. Plankton and algae, for example, degrade water quality when found in excessive quantities because they cause unpleasant odor and taste, turbidity, and reduced oxygen concentrations.

A major water-quality concern is the presence of pathogens, disease-causing organisms including bacteria, viruses and protozoa. Pathogens can cause illnesses such as hepatitis, gastroenteritis and typhoid.

Sources of pathogens include treated and untreated sewage, animal wastes and stormwater runoff. Because it is difficult to measure each of the myriad types of pathogens, certain indicator groups are monitored — most commonly, total coliform, fecal coliform, and enterococci bacteria, which suggest the potential presence of harmful bacteria and viruses.

Pathogens are of special concern in rivers used for drinking water or swimming. This is why sewage discharges are not allowed into rivers that are used as public drinking water sources in Connecticut, and why swimming is prohibited in waters with more than a minimal number of indicator bacteria. Fortunately, bacteria have relatively short life spans in aquatic systems; most die within a few days.

Because edible shellfish such as oysters, mussels and clams filter water through their systems, pathogens can become concentrated in their tissues. Therefore, tidal rivers and coastal waters are closely monitored for the presence of pathogens, and shellfishing is restricted or prohibited when bacteria exceed minimal concentrations.

Toxic pollutants are substances, primarily from industrial and agricultural operations, that have an acute or chronic effect on aquatic organisms, affecting their survival, growth, or reproduction. Many such pollutants accumulate in organisms and sediment over long time periods. The Connecticut Water Quality Standards establish specific criteria for more than 100 toxic compounds, and the state tests acute and chronic water toxicity by monitoring the survival rate of selected aquatic species in water and effluent samples. (See Table 41.)

State waterways that were, or in some cases still are, significantly polluted by toxins include the Naugatuck, Pequabuck and Quinnipiac rivers.

TABLE 41

TYPICAL TOXINS

Metals	Copper, zinc, chromium, lead, nickel, cadmium, mercury
Industrial chemicals	Solvents, chlorinated hydrocarbons, PCBs, cyanide
Hydrocarbons	Gasoline, benzene, phenol, oil
Agricultural chemicals	Pesticides, insecticides, herbicides
Municipal wastes	Chlorine, ammonia

RIVERS AND ASSIMILATION OF WASTE

Waterways play a vital role in the transportation, dilution and assimilation of waste and treated wastewater effluents. Fortunately, natural processes help to cleanse the water of most foreign materials, reducing both the quantity and concentration of pollutants.

TABLE 42

HOW RIVERS REDUCE WASTE

- Allow bacteria and viruses to expire
- Cause decay of organic material
- Transform chemicals into less harmful forms
- Absorb contaminants
- Cause contaminants to settle
- Dilute contaminants
- Aerate and evaporate pollutants

Microscopic organisms in the water aid in the gradual decomposition of organic wastes. However, the decomposition draws oxygen from the water and is satisfactory only as long as oxygen levels are sufficiently high. If there is not enough oxygen, the decomposition causes foul odors and impairs or kills fish and other aquatic life. Thus, it is important to reduce the biochemical oxygen demand of an effluent before it is discharged into a river.

Rivers also dilute materials that do not readily decompose. Although these contaminants may be toxic in high concentrations, they usually are rendered less harmful by dilution. Natural chemical and biological processes in rivers also help to transform some chemical contaminants into less harmful forms. Common industrial wastes include metals, acids and petroleum products.

Swift-flowing rivers have several advantages over sluggish ones. Besides raising oxygen levels, the turbulent flow of swift rivers makes them less likely to trap deposits of contaminated sediments. Turbulence also allows some pollutants to evaporate, removing them from the aquatic environment and diluting them further in the atmosphere. The Naugatuck River, Willimantic River and portions of the Farmington River are examples of watercourses where high flow velocities have helped improve water quality.

Low-gradient rivers with quiet, low flows do not aerate the water well and encourage sediment deposition. This reduces their ability to assimilate waste. Sluggish watercourses include the Still River in Danbury and Brookfield, and the Connecticut, Thames and Quinnipiac rivers.

POLLUTION TRANSPORT

Pollutants are transported by water in several different forms, depending on their solubility and mass.

Dissolved pollutants are soluble materials that mix with water on the molecular level, forming a solution. They include salts, phosphorus and nitrogen, and some metals. Because they do not settle as sediments do and are too small to be filtered out, they are difficult to remove. However, some can be treated through chemical and biological processes.

Suspended pollutants are particulate materials that float or are held within the water, especially by turbulent flow. They represent a large part of the total pollutant load.

Bedload sediments are particulate materials that are pushed along the bottom of the channel by the flowing water. This material usually consists of heavy particles such as sand and gravel and their attached pollutants.

TYPES OF DISCHARGES

Pollutants are discharged either at concentrated sources, known as point sources, or from throughout a general area, when they are termed nonpoint sources. Point sources include sewage treatment plants, treated industrial process or cooling water discharges, and combined sewer overflows.

Nonpoint sources include urban stormwater runoff, agricultural runoff, road drainage, eroded sediments, and leachate from landfills and residential septic systems. These pollutants emanate from a great many dispersed sources and are often conveyed to rivers during rainstorms.

Point sources have traditionally been the most damaging. In Connecticut, point sources are regulated by the Department of Environmental Protection and most meet strict effluent quality standards. Most of the progress made to date is associated with improved treatment and pollution prevention associated with point sources.

Programs have been developed to address several kinds of nonpoint sources as well. The U.S. Environmental Protection Agency has compared the pollution loads from urban runoff with those of raw and treated sewage. (See Table 43.) Note that although pollutants in urban runoff are diluted by rainfall, the total annual loads they contribute to a body of water can be substantial.

POINT SOURCES OF POLLUTION

Waste materials entering rivers from discrete, identifiable locations have been subject to regulation since the passage of Connecticut's Clean Water Act in 1967. The law, which established treatment standards for all sewage and industrial discharges, has proved effective in reducing pollution and improving water quality.

Several thousand municipal and industrial facilities are subject to Department of Environmental Protection discharge permits. More than 900 state pretreatment permits are in effect, requiring commercial and industrial facilities to treat their effluent before discharging it into public sewer systems.

TABLE 43

URBAN RUNOFF VERSUS SEWAGE CONCENTRATION

(All values mg/L except fecal coliform)

CONSTITUENT	URBAN RUNOFF TYPICAL (MEAN) CONCENTRATION	RAW SEWAGE CONCENTRATION	SANITARY SEWAGE EFFLUENT CONCENTRATION
Chemical Oxygen Demand	75	500	80
Total Suspended Solids	150	220	20
Total Phosphorus	0.36	8	2
Total Nitrogen	2	40	30
Lead	0.18	0.10	0.05
Copper	0.05	0.22	0.03
Zinc	0.20	0.28	0.08
Fecal Coliform (organisms per 100 ml)	Up to 50,000	100,000	200

Note: Nonpoint sources contribute significant quantities of organic material, sediments and nutrients to Connecticut's surface water resources. We know pathogens are found in raw sewage. Less is known about the pathogenic resources in urban runoff.

Data Source: United States Environmental Protection Agency, National Urban Runoff Program (NURP) Final Report, 1989

A large wastewater treatment plant used to cleanse sewage discharges.

Combined Sewers

During the 19th and early 20th centuries, a number of communities built what are known as combined sewers, which conveyed both sewage and stormwater runoff directly to rivers. Today, they convey the combined sewage and stormwater to treatment plants. However, during periods of heavy rain, the flow rates can be so high that treatment plants cannot handle the volume and excess flows are discharged to rivers with little or no treatment.

Combined sewer systems are still used in Hartford, New Haven, Bridgeport, Norwich, and Waterbury. Extensive projects have been undertaken to eliminate rainwater from the sewer systems and to treat overflows before discharge. It may still take a decade or two before combined outflows are eliminated.

Wastewater Treatment Plants

Wastewater treatment plants include state, municipal and private facilities. They operate at various levels of efficiency, depending on the characteristics of the raw sewage, plant capacity, and the treatment measures that are used.

Primary sewage treatment includes processes that remove a high percentage of suspended solids and some organic material through measures such as screening and sedimentation tanks. Primary treatment removes up to 35 percent of the material's biochemical oxygen demand and half of the solids.

Secondary treatment includes steps that degrade organic material, helping to further reduce the concentration of pollutants in the effluent. Methods include trickling filters, activated sludge, and aeration lagoons that result in removal of up to 85 percent of both biochemical oxygen demand and solids.

Tertiary, or advanced, treatment refers to a wide range of measures that remove nutrients and toxic material (mostly ammonia) in addition to controlling oxygen demand. Processes include biological treatment, chemical treatment, filtration, and activated carbon. New techniques are being implemented to reduce or eliminate chlorine levels in wastewater effluent, thereby minimizing its toxicity.

Industrial treatment facilities vary with the type of wastewater. Processes may consist of chemical flocculation, sedimentation, or filtration. Organic contaminants can be treated by biological or chemical methods.

Municipal sewage treatment plants operate at various levels of efficiency. The federal Clean Water Act requires all such plants to provide at least secondary treatment. In Connecticut that was achieved in the early 1980s. Numerous facilities now provide more advanced treatment technologies. Since the early '80s many secondary plants have been upgraded and new, advanced plants have been built in Winsted, Wallingford, Bristol, Torrington, Meriden, New Haven, Danbury, Vernon, and Manchester. Also, more than a dozen sewage treatment plants discharging into Long Island Sound or its tributaries have been improved to provide additional removal of nitrogen. The next phase of advanced municipal sewage treatment plant construction will focus on nitrogen removal to benefit coastal waters.

In recent years, more efforts have been made to further reduce contaminant levels, with emphasis on the reduction of nitrogen and chlorine. When improvements were made at the Bristol sewage treatment plant, for example, there was a 90 percent reduction in the levels of biochemical oxygen demand, ammonia, and total suspended solids in the Pequabuck River. Aesthetic quality has improved: The water is clearer, and odor has diminished.

Waste Load Allocations

The Department of Environmental Protection has evaluated many rivers to determine the maximum waste load they can assimilate while still meeting water quality standards. This is done to coordinate the total pollutant load that can be allowed from multiple discharge sources.

Computer models have been developed for many rivers, coastal harbors and Long Island Sound and are an important tool for evaluating water quality. They consider the worst-case condition and determine the maximum sewage discharges allowable during times of warm temperatures and minimum streamflow, when a water's capacity to accept pollutants without reducing water quality is lowest.

NONPOINT SOURCES OF POLLUTION

Nonpoint pollution is that which comes from diffuse, small, intermittent, or mobile sources. While the impact of an individual source

TABLE 44

COMMON NONPOINT SOURCES OF POLLUTION

- Road sand
- Road salt (sodium chloride)
- Motor vehicles (gasoline, oil, grease, metals, rubber)
- Fertilizer
- Pesticides, herbicides, fungicides
- Litter (paper, plastics, foam products, cans, bottles)
- Decayed vegetation (leaves, grass clippings)
- Accidental spillage of petroleum products and other chemicals
- Agricultural fields, manure
- Erosion from construction sites
- Natural soil erosion
- Atmospheric deposition
- Household and industrial solvents and chemicals

Curbs, catch basins, and drainpipes all concentrate surface runoff and reduce infiltration.

TABLE 45

ANNUAL SOIL EROSION RATES WITH VARIOUS LAND USES – UNITED STATES

LAND USE	EROSION RATE (TONS PER SQUARE MILE)	EROSION RATE (RELATIVE TO FOREST = 1)
Forest	24	1
Grassland	240	10
Urban	240	10
Abandoned surface mines	2,400	100
Cropland	4,800	200
Forest clearcut	12,000	500
Active surface mines	48,000	2,000
Construction	48,000	2,000

or rainfall-runoff event often is small, research has found that the cumulative impact is significant. Nonpoint sources include stormwater runoff, atmospheric pollutants, soil erosion, agricultural products, septic systems, accidental spills, and improper disposal of waste materials.

Stormwater runoff often carries pollutants that degrade surface and groundwater resources. Pollutants in stormwater runoff consist largely of materials that accumulate on paved surfaces during dry periods and are washed into watercourses by rainfall.

The type and concentration of pollutants vary with land use, precipitation patterns, development density, vehicular traffic, and control measures. Areas with a high proportion of impervious surfaces and motor vehicles have relatively many sources of pollutants and allow little water infiltration. Thus they generate a larger share of both runoff and contaminants.

Because the transport of pollutants from most nonpoint sources depends on rainfall frequency, duration, and intensity, they have a discharge pattern that is highly variable.

Although urban nonpoint pollution is diluted by rainfall, the total annual loads are considerable. (See Table 43.)

Soil Erosion

Soil erosion is one of the major sources of nonpoint pollutants. While erosion is a natural process, it is greatly accelerated by urbanization, resulting in unnaturally high sediment loads in many watercourses.

Beginning in the mid 1970s and continuing through the '80s, the U.S. Environmental Protection Agency encouraged the inventory of erosion and sediment conditions. The U.S. Natural Resources Conservation Service (formerly the U.S. Soil Conservation Service) became more involved in controlling erosion in areas undergoing development and not just in agricultural areas, its traditional area of involvement.

In Connecticut, the design of most storm-drainage systems for highways and land development now incorporates erosion-control measures. The Connecticut office of the Natural Resources Conservation Service has published erosion-control handbooks since 1972 and held training seminars. In 1974, the County Soil and Water Conservation Districts in Connecticut were reorganized to cooperate with the Department of Environmental Protection. The passage of the Connecticut inalnd wetland protection acts provided significant permitting and enforcement authorities for improving sediment and erosion control.

State legislation in 1977 that enabled local planning and zoning boards to require erosion-control regulations was strengthened in 1983 with Public Act 83-388. This law mandated that municipalities adopt erosion-control regulations by July 1, 1985, and required soil-erosion and sediment-control plans for most land development to help protect water quality.

When vigorously enforced, measures to control erosion and sedimentation during development can be effective in minimizing runoff-quality problems.

Many erosion-control measures, such as hay-bale barriers and silt-containment fences, have only a short-term benefit during the construction period itself. Other measures, such as stable channel banks, vegetated ground covers, and sediment basins, have long-term benefits.

Stormwater Runoff Quality

The impact of nonpoint sources is evident in both urban and agricultural areas. Many urban streams have limited recreational value and wildlife habitat because of poor water quality resulting from street and land runoff and thermal pollution. Streams in agricultural areas often suffer from high sediment loads and excess nutrients.

Urban runoff introduces significant pollutants. Common pollution sources are automobile products, de-icing materials, asphalt products, airborne materials, and spillage.

Many of the pollutants come from roadways and parking lots. The Environmental Protection Agency has found an average of 1,400 pounds of loose material on each mile of roadway in urban areas. Industrial areas have the highest amounts; central business districts have the lowest. Seventy-eight percent of this material is located within six inches of the curb.

In its most comprehensive assessment of the impact of urban runoff, the Environmental Protection Agency, in the early 1980s, sampled 78 sites in 28 cities over two years. The study found that runoff quality is not just a function of rainfall intensity or depth, but is highly variable and directly related to how the land is used.

The results (shown in Table 47) indicate that heavy metals are a

TABLE 46

THE POLLUTANTS IN STORMWATER RUNOFF

POLLUTANT	WHERE IT COMES FROM
Asbestos	Clutch plates, brake linings
Bacteria	Pets, animals, birds, septic systems, soil bacteria
Cadmium	Tire wear (filler material), insecticide application
Chromium	Metal plating, moving engine parts, brake-lining wear, crankshafts, piston rings
Copper	Metal plating, bearing and bushing wear, moving engine parts, brake-lining wear, fungicides and insecticides
Cyanide	Anti-cake compound (ferric ferrocyanide, sodium ferrocyanide, yellow prussate of soda) used to keep de-icing salt granular
Hydrocarbons	Oil, paraffin lubricants, antifreeze, hydraulic fluids
Iron	Auto body rust, steel highway structures (guardrails, etc.), moving engine parts, soil erosion
Lead	Tire wear (lead oxide filler material), lubricating oil and grease, bearing wear, paint
Manganese	Moving engine parts, soil erosion
Nickel	Diesel fuel and gasoline (exhaust), lubricating oil, metal plating, bushing wear, brake lining wear, asphalt paving
Nitrogen, phosphorus	Atmosphere, fertilizer application, motor oil additives, decomposing plants, root material, grass clippings, animals, septic systems, paper, exhaust emissions
Organic chemicals	Agricultural and lawn-care pesticides, herbicides, transformers, hydraulic fluids
PCBs	Spraying of highway rights-of-way, background atmospheric deposition, PCB catalyst in synthetic tires
Particulates	Pavement wear, vehicles, atmosphere, road maintenance, soil erosion, channel erosion, traction-control materials
Petroleum	Spills, leaks or blow-by of motor lubricants, antifreeze and hydraulic fluids, asphalt surface leachate
Rubber	Tire wear
Sodium, calcium, chloride	De-icing salts
Sulphate, sulphur	Roadway beds, de-icing salts, fossil fuels, road sand
Suspended solids	Soil erosion, pavement wear, litter, leaves, fuels
Zinc	Tire wear (filler material), motor oil (stabilizing additive), grease

TABLE 47

CONNECTICUT INDUSTRIAL STORMWATER MONITORING RESULTS, 1996

Parameter	Performance Criteria	Units	25%	50%	75%	90%	95%	MAX
			Percentile					
Oil and grease	20	mg/L	1.0	2.4	5.0	7.0	13.0	332.0
Chemical oxygen demand	50	mg/L	15.0	32.1	71.3	133.5	205.2	7,000
Total suspended solids	50	mg/L	7.0	20.0	64.8	200.0	384.9	18,087
Total phosphoric	0.5	mg/L	0.06	0.13	0.31	0.69	1.03	67.10
Total K nitrogen	3	mg/L	0.60	1.10	2.10	3.90	5.61	93.50
Nitrates	3	mg/L	0.26	0.57	1.12	2.19	3.12	60.00
Fecal coliform bacteria	2,000	#100ml	10	100	1,500	10,000	11,000	1,023,000
Copper	0.2	mg/l	0.01	0.04	0.08	0.20	0.40	73.37
Lead	0.05	mg/L	0.01	0.02	0.05	0.09	0.16	21.70
Zinc	0.5	mg/L	0.05	0.14	0.32	0.75	1.18	63.60

Source: CT DEP Water Bureau, 1996

primary problem in urban runoff, often exceeding EPA water quality standards. Concentrations frequently exceeded the levels above which they could be expected to cause long-term harm to plants or animals. In addition, levels of coliform bacteria in runoff were very high during storms, often exceeding levels permitted in drinking water.

Residential areas contribute pollutants from many sources. Lawns and gardens provide fertilizers, herbicides, insecticides, and leaf detritus. On-site sewage disposal systems with septic tanks and leaching systems discharge partially renovated wastewater into the soil, some of which may subsequently affect surface water quality. Other sources include household materials such as solvents, paint remover, fuel oil, lead paint, and copper roofing materials.

Rural areas generate natural pollutants, including products of erosion, plus nitrogen and phosphorus from decaying vegetation.

Runoff from agricultural areas can contain greater quantities of such natural nutrients, in addition to fertilizers, manure, bacteria, viruses, etc.

While not as easily transported by runoff, some types of litter may enter watercourses at a greater rate during storms. Litter can include anything from grass clippings to shopping carts or large household appliances. Some litter materials — such as plastics, glass and metals — do not decay readily, remaining in streams for long periods and degrading aesthetics. Some of this waste, such as broken glass or rusted metal, can injure people who use rivers for recreation.

De-icing Salts and Aggregate

During winter, salt and sand mixtures are applied to roads to melt snow and improve traction. Application rates can be as much as several tons per mile per year. Salt is also used to prevent sand stockpiles from freezing. The most common form, sodium chloride, can create health problems. It also harms roadside vegetation and accelerates corrosion of metals and concrete, including vehicles, roads, and bridges.

Runoff from snowmelt and rainfall conveys road sand and salt. Since salt dissolves in water, it cannot be easily removed from road runoff. Fortunately, the salt carried to streams and rivers is usually diluted to low concentrations and seldom impacts aquatic life. However, areas where a lot of salt is used or stored can have excessive concentrations in groundwater. Sodium concentrations in wells frequently exceed the

TABLE 48

RELATIVE IMPORTANCE ASSIGNED TO NONPOINT SOURCE POLLUTION CATEGORIES

NONPOINT SOURCE CATEGORY	LAKES	RESERVOIRS	RIVERS & STREAMS	ESTUARIES	INLAND WETLANDS	TIDAL WETLANDS	GROUND WATER
Boats and marinas	—	—	—	Mod.	—	Low	—
Agriculture	High	High	Low	Low	Mod.	—	High
Silviculture	Low	Low	Low	—	Low	—	—
Construction	Mod.	High	Mod.	—	High	High	—
Land development	High	High	Mod.	Mod.	High	High	High
Urban runoff	Mod.	High	Low	High	Mod	High	—
Resource extraction	—	—	Mod.	Mod.	—	—	—
Solid waste disposal	High	Mod.	Mod.	Mod.	Low	Low	High
Hydrologic/habitat modification	—	—	Mod.	—	Mod.	High	—
Material storage/leaks	Low	High	Low	—	Low	Low	High
Atmospheric deposition	Low	Low	Low	High	—	—	—
Highway maintenance and runoff	Mod.	Mod.	Low	Low	Low	Low	Mod.
Transportation-related spills	Low	Mod.	Mod.	Mod.	Mod.	Mod.	Mod.
Industrial site contamination	Low	Low	Mod.	Mod.	Mod.	Mod.	High
Natural	Low	Low	—	—	—	—	—
Waterfowl	Low	Low	—	—	—	—	—

Mod. = moderate

Source: CT DEP

state drinking water recommendation of 28 milligrams per liter.

The sand aggregate applied to roads for traction control is frequently transported by snowmelt and runoff into storm drains and streams. This sand is a significant source of sediment and often buries natural streambeds, where it can smother the aquatic biologic community. Ponds, lakes and wetlands may suffer similar habitat impacts. Recommended control methods include street sweeping, use of deep sumps at catch basins and drainage inlets, and use of large underground grit chambers along storm drains above their discharge points.

Several de-icing chemicals have been tested as substitutes for sodium chloride. Calcium chloride is sometimes used as an alternative in sensitive watershed and aquifer areas. It is particularly effective at temperatures below 20 degrees Fahrenheit. However, it costs several times as much as sodium chloride. Other alternatives are in the testing stage.

The Connecticut Department of Transportation is responding to the concern about salt by reducing application rates, modifying application practices, and constructing new sheds to store salt supplies.

Thermal Pollution

The unnatural change in water temperature resulting from human activities is called thermal pollution. Because aquatic species are sensitive to temperature and because warm water can hold less dissolved oxygen, thermal pollution can force fish to relocate, alter migration patterns, and even cause fish kills. Causes of thermal pollution include discharges of water that has been used for industrial cooling, thermal and nuclear power plants, urban runoff, and removal of riparian vegetation leading to increased solar exposure.

The effect that thermal pollution resulting from urbanization and runoff has on local streams has been well documented in Maryland. A study in that state found that urban runoff and land uses can raise the water temperatures in streams by as much as 20 degrees Fahrenheit. This can have a significant effect on cold-water species: The numbers of cold-water algae and insects decline severely, and warm temperatures are potentially fatal to fish such as trout. The study also determined that the temperature of stormwater runoff contained by detention and retention basins increased as much as nine degrees. Connecticut, as do other states, has a permit program regulating thermal discharges.

The importance of riparian vegetation as a thermal insulator of small, headwater streams must not be underemphasized. During summer especially, an umbrella of dense vegetation typically shields natural streams from solar radiation, thereby reducing daily temperature fluctuations. The common development practice of removing riparian vegetation can greatly alter the temperature regime of small streams.

Stormwater Runoff Quality Control

The basic approach to maintaining the quality of stormwater runoff emphasizes multiple techniques to control pollutants at their source, control the volume and rate of runoff, and then treat the remaining runoff.

The practice in many states, including Connecticut, is to employ Best Management Practices that prevent or reduce the discharge of pollutants to surface waters and groundwater. The intent is to limit the use of hazardous materials, limit the concentration of pollutants and minimize their entry into drainage systems.

Occasionally, an alternate approach is used that attempts to apply to stormwater runoff the same standards that mandate the quality of discharges of treated sewage effluent. These standards are difficult to apply, however, because of the variation in flow rates of surface runoff and in the concentrations of pollutants it carries.

Below are examples of practices used to control pollutants at the source. They range from conservation and prudent use of hazardous materials to regulation and physical controls.

The use of runoff quality control measures has become common, especially in communities whose residents are sensitive to environmental concerns. Local land use commissions can help control new nonpoint sources of pollution by requiring that water quality be addressed before issuing site development permits.

The Connecticut Department of Environmental Protection is also increasingly involved through Section 402 of the Clean Water Act, the Environmental Protection Agency's National Pollutant Discharge Elimination System. Published in 1990, these regulations emphasize the use of Best Management Practices and require industries and large cities to obtain federal permits for existing and new stormwater discharges.

Planting grass around this sediment basin would reduce the amount of sediments reaching it, and a border of trees would reduce thermal pollution of its outflow.

A sediment basin with appropriate landscaping.

Stormwater Runoff Collection Systems

Use of low-maintenance, artificial drainage systems has increased the transport of pollutants to watercourses. Most modern drainage systems are designed to collect and convey surface runoff from developed areas to discharge points as quickly as possible and to be self-cleaning. They also tend to collect and concentrate pollutants in runoff and discharge it directly into watercourses with limited opportunities for settlement or filtration. Such high-impact drainage systems include:

- Curbed, paved surfaces that concentrate runoff on the pavement, minimizing soil infiltration.
- Drainage inlets that direct surface runoff into underground pipes.
- Storm drain pipes that carry roadway and parking lot runoff, with its pollutants, to watercourses.
- Direct discharges.

The direct discharge of stormwater into watercourses should be avoided wherever possible. Innovative drainage systems that accomplish this minimize the collection of pollutants and provide multiple opportunities to trap and treat them before the runoff is discharged. Such systems:

- Minimize impervious ground cover.
- Have discontinuous pavements, using grass shoulders and vegetated islands to distribute sheetflow to pervious surfaces.
- Lack curbing, thus encouraging overland flow.
- Employ grass swales to encourage settlement and filtration of pollutants.
- Use sediment basins to trap floatable objects and materials that can settle out.
- Include infiltration devices to reduce runoff rates and volumes.
- Use wet detention basins (those that include a vegetated natural or man-made pond possibly with an artificial wetlands component) to reduce the volume of runoff and improve its quality through natural processes.

All development projects should be required to provide emergency action plans for use during accidents or spills. These plans should include written instructions, communication procedures, and provisions for safe storage of sources of potential pollutants. New commercial and industrial sites should be required to provide means for containing and trapping oil, gasoline, and chemicals, and existing facilities should be retrofitted to the same end.

While urban flood control systems are designed for large, rare rainfall events, measures to control the water quality of urban runoff should be proportioned for smaller and more frequent storms. This is important because most pollutants in urban runoff are generated continuously and are conveyed into water bodies by routine rain events. The typical recommended criteria call for controlling the quality of the first inch of watershed runoff. This approach is effective because most storms produce less than one inch of runoff, and pollutants tend to be concentrated in the initial runoff.

BENTHIC SEDIMENT DEPOSITS

Sediments in the riverbed, known as benthic deposits, also influence water quality.

In some water bodies, particularly ponds and streams with low flow velocities, such sediments may contain accumulations of organic material — from natural or human sources — that has settled out of the water. These include:

- Urban and agricultural runoff
- Sludge from treated sewage effluent
- Industrial discharge sludge
- Leaves and woody detritus
- Phytoplankton
- Waste from construction sites
- Aggregate from road sanding

Decomposition of organic benthic sediments and the aquatic plants whose growth they stimulate uses oxygen. If available dissolved oxygen is depleted beyond a certain point, decomposition becomes anaerobic, occurring without oxygen. This causes the release of hydrogen sulfide and carbon dioxide in gas bubbles that float sediments to the surface. It creates odors and alters the biological species present. For healthy, aerobic decomposition, one to five grams per day of oxygen is required for each square meter of streambed.

Some microorganisms, including coliform bacteria, adhere to and settle with sediments to the streambed, where they appear to have a longer life than in open water. Such organisms are periodically resuspended by high runoff, recontaminating the water.

TABLE 49

NONPOINT POLLUTION CONTROL PROCESS

- Identify receiving water resources
- Review regulatory requirements
- Set goals and priorities
- Delineate watershed
- Address hydrological issues
  - Maintain to the extent possible the natural drainage pattern
  - Design for frequent runoff events
  - Determine runoff rates and volumes
  - Try to reduce runoff rates and volumes
  - Encourage infiltration
  - Minimize connected impervious cover
- Identify potential sources of pollution
- Control pollutants at their source
- Use multiple Best Management Practices
  - Remove settleable solids
  - Trap floating materials, litter and oil
  - Use grass and other vegetated filter zones
  - Employ long-duration detention
  - Use wet detention basins rather than dry ones
- Identify potentially unavoidable impacts and address methods to minimize them
- Prepare written operation and maintenance plan

The accumulation of contaminants that do not decay readily, including heavy metals, is also a problem that can cause chronic or acute toxicity. They concentrate in the sediment layer, and can be resuspended and carried downstream by floodwaters. In rare, acute cases of sediment contamination, the only recourse is to seal it in place or remove it by dredging.

The Housatonic River has one of the largest benthic sediment problems in Connecticut. Over 100 miles of river bottom in Connecticut and Massachusetts has been contaminated by PCBs (polychlorinated biphenyls), a heavy chlorinated organic fluid, which concentrates in fish tissue. Public-health advisories have been issued in both states as a result. Other water bodies with contaminated sediments include the Versailles Pond on the Little River in Sprague, the Mill River in Fairfield and major harbors along the coast.

The presence of contaminated sediments in riverine impoundments created by dams is a problem that complicates removal of old, obsolete structures.

IMPACT OF RUNOFF ON LAKES AND RESERVOIRS

Additional factors affect water quality in lakes and reservoirs because of their poor circulation and the existence of different temperature strata, each with its own biological community. The long period during which water is held and the lack of significant flow velocity results in increased settling of solids, reduced aeration and oxygen levels, and increased solar exposure and biological production. All of these, in turn, are affected by stormwater runoff and pollution conveyed by tributaries and direct overland flow.

Causes of lake degradation include nutrient enrichment, oxygen depletion, and siltation.

All lakes and ponds undergo a slow, natural, three-stage aging process, called eutrophication, that can be accelerated by human activities. Oligotrophic lakes are those in the earliest stage. They are deep with clear, infertile waters, little sediment accumulation on the bottom, and little or no algae and aquatic weed growth. Even the deepest water is well oxygenated. Eutrophic lakes are in the latter stages of the aging process, characterized by significant sediment accumulation, abundant algae growth and low oxygen levels. By this stage, a lake is less appealing for recreation or water supply. Mesotrophic lakes are in the transition stage, with moderate clarity, levels of dissolved oxygen and algae growth.

Eutrophication is the predominant concern relating to water quality in Connecticut's lakes. Excessive growth of algae and aquatic plants is the primary factor that impairs recreation. An important secondary concern is depletion of dissolved oxygen during the summer and the resulting degradation of habitat for cold-water fish species. Contamination by sewage-borne pathogens is a potential problem, but is rarely encountered in Connecticut's recreational lakes.

The accumulation of pollutants and algae biomass is more severe in lakes with a long hydraulic exchange period: the time it takes water to pass through the lake. Algae and pollutants have less opportunity to accumulate in lakes with a short exchange period (10 days or less) because they are quickly transported through the system.

Since point discharges of pollution are prohibited in reservoir watersheds, the largest threat to these sources of drinking water is the cumulative effect of nonpoint pollution.

Lake rehabilitation in process — weed harvesting.

Lake Management

The general approach to lake management in Connecticut focuses on controlling the undesirable aspects of eutrophication — algae blooms, low oxygen levels, high turbidity, and excessive rooted plants — as well as high bacteria counts.

The most important step in controlling eutrophication is to implement watershed management practices that reduce the input of nutrients and sediments.

Some techniques can be implemented at the lake itself. These include increasing water circulation, harvesting aquatic weeds, chemically treating the water to kill algae, dredging bottom sediments, and drawing down water in winter to permit removal of sediments and to kill normally submerged aquatic plants.

Major sources of phosphorus reaching a lake in runoff include septic system effluent, fertilizer, leaves, animal wastes, urban runoff, and nutrient-rich sediments. All of these are best controlled at their sources, with emphasis on repairing septic systems and avoiding soil erosion and excessive use of fertilizer.

The excess turbidity that reduces water clarity in ponds results from algae and from dissolved and suspended materials, such as silt and clay. These are also most easily controlled at their source, for example through land management practices that reduce erosion.

TABLE 50

LAKE MANAGEMENT METHODS

- Control septic systems
- Limit fertilizer use
- Circulate and aerate artifically
- Control soil erosion
- Harvest weeds
- Control weeds chemically
- Dredge bottom sediments
- Draw down in winter

TABLE 51

LAKESHORE RUNOFF QUALITY CONTROL METHODS

- Preserve wetlands in lake watersheds
- Preserve riparian buffers
- Control surface soil erosion
- Provide channel erosion control
- Trap sediment
- Provide biological filters to consume nutrients

Lake management requires management of nonpoint sources to prevent pollutants from reaching rivers and wetlands that are tributary to the lake.

The U.S. Environmental Protection Agency and many state agencies have published detailed manuals on lake restoration techniques that emphasize the importance of watershed management. One such manual is *Caring for Our Lakes*, published by the Connecticut Department of Environmental Protection. This manual identifies the following as important techniques to control the water quality of runoff: erosion control, detention basins, sediment basins, grass channels, and protection of natural riparian zones. The manual also emphasizes the need for careful planning, construction and maintenance of septic systems.

STREAMBANK RIPARIAN AREAS (Buffers)

Even where there is no direct discharge into a stream, local land uses have a strong influence on water quality. For example, simply removing vegetation along a river bank can:

- Reduce shade, increasing water temperature.
- Reduce oxygen levels because of the higher water temperature.
- Reduce root mass, increasing bank erosion and, thus, sediment levels.
- Reduce filtration of runoff.
- Eliminate wildlife cover and food sources.
- Reduce organic detritus.

The Department of Environmental Protection Fisheries Division recommends that a 100-foot-wide vegetated buffer zone be left along perennial watercourses to help protect water quality and aquatic life. (See Chapter Seven.)

BIOLOGICAL HABITAT ASSESSMENTS

Increasingly, biological methods are being used to evaluate both water quality and the effects of various activities on aquatic habitats.

The procedure uses the pollution tolerance of selected aquatic insects as an indicator of water quality. Insects such as mayflies, caddisflies and stoneflies have a low tolerance for pollution and their presence implies good water quality. Other invertebrates, such as certain benthic worms, leeches, and midge flies, have a higher tolerance for pollution.

The U.S. Environmental Protection Agency Rapid Bioassessment Protocol (RBP) uses the ratio between pollution tolerant and intolerant species to rate water quality. Such techniques are advantageous because they are cost-effective and assess the cumulative stress of pollutants on the aquatic ecosystem. However, they do not identify specific pollutants and therefore do not help determine the source of contaminants or possible corrective measures.

The second stage of the RBP evaluates physical habitat conditions to assess the overall quality of the aquatic habitat. It compares the water quality and habitat characteristics with the conditions preferred by individual species.

The Instream Flow Incremental Methodology (IFIM) extends the concept of biological habitat evaluations to forecast the impact of flow diversions, channelization, and habitat improvements. It is based on the correlation of fish species and populations to water quality, depth, substrate, velocity, and cover. The above data are obtained for a series of incremental flow rates, either natural or artificially regulated, to assess the effects on habitat size and quality. The same approach has been used to evaluate the effect of variable flow rates on a stream's ability to assimilate wastewater, on its aesthetics, and on its potential for recreation.

LONG ISLAND SOUND STUDY

Connecticut, in cooperation with New York and the federal government, conducted a study to address water quality problems in Long Island Sound. The findings revealed that wastes from sewage treatment plants and nonpoint sources cause low oxygen levels in the Sound during the summer and reduce fish and shellfish populations.

The principal cause is believed to be excessive nitrogen, which stimulates the growth of plankton. After the plankton die and sink to the bottom, their decay consumes oxygen.

In September 1994, Connecticut and New York approved a comprehensive plan for managing Long Island Sound, with primary emphasis on improving water quality. The six areas to receive special attention are:

- Low dissolved oxygen
- Toxic contamination
- Pathogen contamination

- Floatable debris
- Fish and shellfish
- Habitat loss

Connecticut is implementing a broad program to address these issues. The Department of Environmental Protection and shoreline municipalities are modifying wastewater treatment plants to reduce the discharge of nitrogen, and no future point source discharges will be allowed directly into the Sound. Another important part of the Long Island Sound cleanup program is the control of nonpoint sources of pollutants that reach rivers and streams, an effort that encompasses their full watersheds.

FUTURE ISSUES

The success of water pollution control programs in Connecticut has been accompanied by increased competition for using the clean waters. When rivers were polluted, there was little interest in using them for fishing, boating, swimming, tubing, water supply, or riverbank development. As the water quality improves, there is increased interest in using rivers for many activities other than waste assimilation. These competing demands create a need to allocate resources wisely for multiple uses.

Further improvements in water quality will require changes in the way we use and dispose of water, with more emphasis on water conservation and controlling pollutants at their source. There also is a tremendous need for the development and use of nontoxic products to replace toxic materials and thereby reduce sources of pollution.

Similarly, expansion of waste recycling programs presents a technical, financial and social challenge that we have only begun to undertake.

Tidal Rivers and Marshes

To stand at the edge of the sea, to sense the ebb and the flow of the tides, to feel the breath of a mist moving over a great salt marsh, to watch the flight of shore birds that have swept up and down the surf lines of the continents for untold thousands of years, to see the running of the old eels and the young shad to the sea, is to have knowledge of things that are as nearly eternal as any earthly life can be.

Rachel Carson, foreword, *Under the Sea-Wind*, 1941

FIGURE 36
SEA LEVEL TERMS & RELATIONSHIPS

FIGURE 36 A

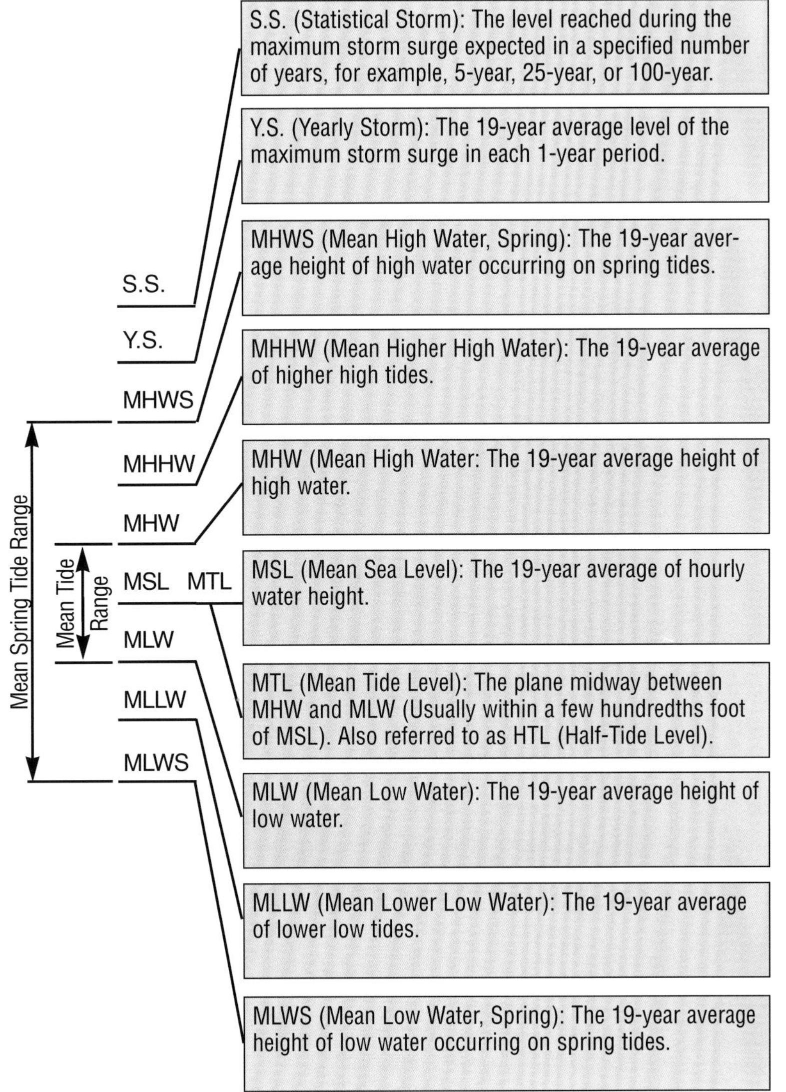

Note: In Connecticut, Mean Sea Level is currently higher than that indicated by the National Geodetic Vertical Datum (NGVD).

Source: Modified from Coastal Ecosystem Management by John Clark, 1993

FIGURE 36 B

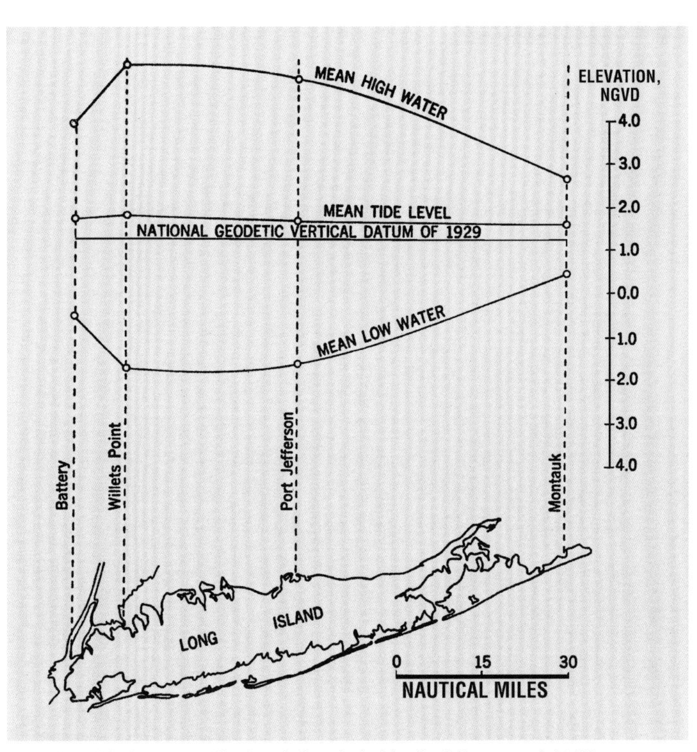

Relationship between National Geodetic Vertical Datum of 1929 and mean tide level, mean high water, and mean low water for tide stations between Montauk and the Battery, NY

Source: National Oceanic and Atmospheric Administration.

Tidal Rivers and Marshes

Long Island Sound and the smaller Fishers Island Sound embrace a 1,300-square-mile estuary whose flow patterns are dominated by ocean tides rather than runoff. Twice a day, tidal waters surge into the Sound, entering primarily from the eastern end and moving west along its 110-mile length. Twice a day they flow back out.

The tides have a significant influence. There is a unique hydrology — they ebb and flow over an approximately 12-hour cycle. The tides bring sea salts to riverine areas and fresh water to saltwater areas.

The mix of fresh and salt water creates rich, diverse ecosystems in tidal rivers and marshes. These critical resources, like the waters of the Sound itself, are subject to the pressures of surface runoff, sewage effluent, and other factors resulting from the high population density of coastal areas.

Salt water from Long Island Sound flows into coastal channels during high tide, raising water elevations until the tide falls again. The downstream portion of tidal-influenced rivers usually is a mix of fresh and salt water, forming riverine estuaries with brackish water. The upstream limit of tidal influence can extend farther inland than the saltwater mixing zone, above which it is termed the tidal river's freshwater segment.

Where some tidal rivers enter the Sound, their partly submerged channels form harbors, as in Norwalk (Norwalk River), Westport (Saugatuck River), Bridgeport (Pootatuck River), New Haven (Quinnipiac and Mill Rivers), and New London (Thames River).

Many coastal rivers, such as Carolina Creek in East Haven and Pine Creek in Fairfield, are located behind barrier beaches or sand bars, draining the area between these formations and the mainland.

TABLE 52
TIDAL LENGTH OF CONNECTICUT'S LARGER RIVERS

RIVER	WATERSHED (SQUARE MILES)	APPROXIMATE TIDAL LENGTH (MILES)	HEAD OF TIDEWATER
Connecticut	11,265	60	Enfield
Housatonic	1,949	12	Derby
Thames	1,473	15	Norwich
Quinnipiac	166	10	Wallingford
Saugatuck	92	3	Westport
Norwalk	64	2	Norwalk

TABLE 53 DEFINITIONS OF COMMON TIDAL TERMS	
Datum	A base elevation used as a reference, from which to determine heights or depths.
High tide line	The highest predicted tide of the year.
Mean high water	A datum based on the average of all high water heights observed over a 19-year period, called the National Tidal Datum Epoch (see below).
Mean low water	A datum based on the average of all low water heights observed over the epoch.
Mean sea level	A datum based on the arithmetic mean of hourly heights observed over the epoch.
Mean tide level	A datum which is the mean of the mean high water and mean low water, also called half-tide level.
National Geodetic Vertical Datum of 1929 (NGVD)	A reference adopted as a standard datum for elevations. This datum is fixed and does not take into account changes in sea level. The relationship between the NGVD and local mean sea level varies with time and from one location to another.
National Tidal Datum Epoch	The specific 19-year period adopted by the National Oceanic and Atmospheric Administration as the official time segment over which tide observations are taken to determine mean values.
Neap tide	The tide that occurs about the time the moon is at its first and third quarters. This tide occurs semi-monthly. It does not rise as high or fall as low as the average.
Range of tide	Difference between high water and low water.
Spring tide	The tide that occurs at or near the time of new and full moon, which rises highest and falls lowest from the mean level. It occurs semi-monthly.

This type of coastal river may not extend into upland areas and may not have much fresh water, its outflow being primarily the sea water brought in at high tide. Such a channel may even be dry at low tide, due to the limited freshwater runoff.

TIDES

Tides in the Northeastern United States are semi-diurnal, with one high tide and one low tide occurring every 12 hours and 25 minutes. They are caused by the gravitational pull of the sun and moon and are closely related to the lunar day.

A rising tide is also called the flood tide; a falling tide is known as the ebb tide. Slack water — a brief period of virtually no flow — occurs at both the peak and trough of the tide cycle.

In Long Island Sound, the mean tide range, the difference in elevation between low water and high water, varies from 2.8 feet at the Rhode Island border to 7.4 feet at the New York border (see Figure 36). The geometry of the Sound contributes to this amplification of the ocean tide.

A spring tide occurs at the new and full moons, when the gravitational force of the moon and sun act together. (See Figure 36.) The most extreme spring tides, both high and low, occur near the spring and autumn equinoxes. The spring tide range in Long Island Sound is typically from 0.5 to 1.8 feet greater than the mean tide range.

Tide levels are also influenced by daily variations in atmospheric pressure, wind speed, and wind direction. On-shore winds from the south or east and low barometric pressure will increase tide levels, whereas off-shore winds and high pressure will reduce them. The two high tides (and low tides) in each lunar day usually do not have equal magnitudes. Thus there is a higher high tide and a lower high tide in each lunar day. The difference between them in the Sound is usually less than one foot.

The mean tide level is the midpoint between mean high water and mean low water. In the Sound, mean tide level is usually about 0.2 feet higher than mean sea level. This difference occurs because although there is a limit to how low the tide can subside, some high tides can be greatly intensified by storms.

Mean sea level is the average elevation of the sea at a particular location, based on hourly measurements observed over a 19-year cycle called the National Tidal Datum Epoch. This is recomputed every 19 years to reflect long-term changes in sea level, and is reviewed annually for possible shorter-term revision. The Connecticut data for the tidal epoch are obtained by the National Oceanic and Atmospheric Administration from two official tide gauges, in New London and Bridgeport.

Sea levels vary on a seasonal as well as long-term basis. Mean sea level, mean high water, and mean low water all tend to rise during the summer and early autumn and decline during the winter. This is attributed to changes in the volume of sea water as its temperature rises and falls, and to slight variations in gravitational pull as the Earth's inclination changes relative to the sun and moon.

The tidal prism refers to the total volume of water that flows in and out of a tidal river or estuary during each tide cycle, or the difference in volume between high tide and low tide. The tidal exchange rate is the percentage of the previous tidal prism that is returned to the riverine estuary during the next tidal cycle; this is an important factor in the dilution of runoff and wastewater.

Tidal Lag

The tides in Long Island Sound and its tributary rivers do not occur simultaneously. The rising tide is a progressive wave that moves westward up the Sound from Stonington to Greenwich. It reaches its peak in Greenwich more than two hours after its peak in Stonington.

Similarly, it takes time for the high tide to move inland along tidal rivers. Thus, the tide cycle of a river usually lags behind that of its inlet. On the Connecticut River, for example, there is a six-hour lag between high tide in Old Saybrook and that of Hartford. The lag in any particular channel depends on its length, depth, width, and roughness.

Table 54 gives a relative schedule of tidal lags for locations on the Connecticut shoreline and some rivers, along with average tide heights and ranges.

Tidal Circulation

Tidal rivers have complex flow patterns. Fresh water flowing downstream encounters salt water being driven inland by rising tides and, at high tide, sea currents. Consequently, rates and direction of flow change constantly.

As the tide rises in the Sound, sea water is pushed into the mouths

TABLE 54

TIDAL DIFFERENCES

Location	DIFFERENCES IN TIME HIGH (HRS MINS.)	DIFFERENCES IN TIME LOW (HRS MINS)	RANGES MEAN (IN FEET)	RANGES SPRING (IN FEET)	MEAN TIDE LEVEL (FT ABOVE MEAN LOW WATER)
RELATIVE TO NEW LONDON					
Stonington, Fishers Island Sound	-0 32	-0 41	2.7	3.2	1.5
Noank, Mystic River Entrance	-0 22	-0 08	2.3	2.7	1.4
Thames River					
New London, State Pier*	0 00	0 00	2.6	3.0	1.5
Smith Cove Entrance	0 00	+0 10	2.5	3.0	1.4
Norwich	+0 13	+0 25	3.0	3.6	1.7
Millstone Point	+0 09	+0 01	2.7	3.2	1.5
Connecticut River					
Saybrook Jetty	+1 11	+1 45	3.5	4.2	2.0
Saybrook Point	+1 11	+0 53	3.2	3.8	1.8
Lyme, Highway Bridge	+1 25	+1 10	3.1	3.7	1.7
Essex	+1 39	+1 38	3.0	3.6	1.7
Hadlyme	+2 19	+2 23	2.7	3.2	1.5
East Haddam	+2 42	+2 53	2.9	3.5	1.6
Haddam	+2 48	+3 05	2.5	3.0	1.4
Higganum Creek	+2 55	+3 25	2.6	3.1	1.5
Portland	+3 51	+4 28	2.2	2.6	1.3
Rocky Hill	+4 44	+5 44	2.0	2.4	1.2
Hartford	+5 30	+6 52	1.9	2.3	1.1
RELATIVE TO BRIDGEPORT					
Westbrook, Duck Island Roads	-0 24	-0 32	4.1	4.7	2.2
Duck Island	-0 26	-0 35	4.5	5.2	2.4
Madison	-0 21	-0 30	4.9	5.6	2.6
Falkner Island	-0 14	-0 25	5.4	6.2	2.9
Sachem Head	-0 11	-0 15	5.4	6.2	2.9
Money Island	-0 12	-0 23	5.6	6.4	3.0
Branford Harbor	-0 08	-0 18	5.9	6.8	3.1
New Haven Harbor Entrance	-0 09	-0 14	6.2	7.1	3.3
New Haven (City Dock)	-0 01	-0 01	6.0	6.9	3.2
Milford Harbor	-0 08	-0 10	6.6	7.6	3.5
Stratford, Housatonic River	+0 26	+1 01	5.5	6.3	2.9
Shelton, Housatonic River	+1 35	+2 44	5.0	5.8	2.7
Bridgeport*	0 00	0 00			
Black Rock Harbor Entrance	-0 04	-0 03	6.9	7.9	3.7
Saugatuck River Entrance	-0 02	+0 01	7.0	8.0	3.8
South Norwalk	+0 09	+0 15	7.1	8.2	3.8
Greens Ledge	-0 02	-0 01	7.2	8.3	3.9
Stamford	+0 03	+0 08	7.2	8.3	3.9
Cos Cob Harbor	+0 05	+0 11	7.2	8.3	3.9
Greenwich	+0 01	+0 01	7.4	8.5	4.0

*Official National Oceanic and Atmospheric Administration tide gauge

FIGURE 37
TIDAL RIVER PROFILES

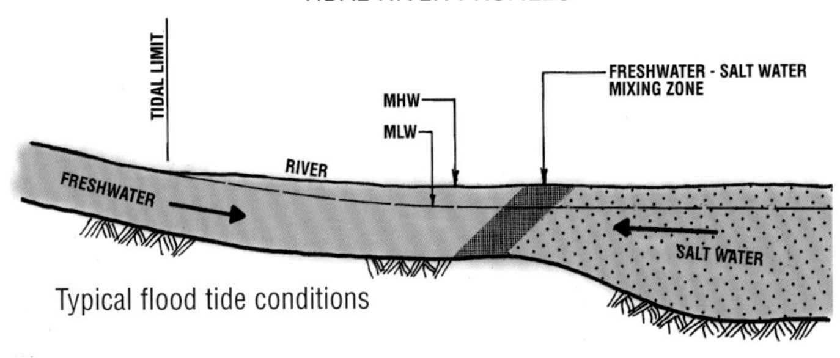

Typical flood tide conditions

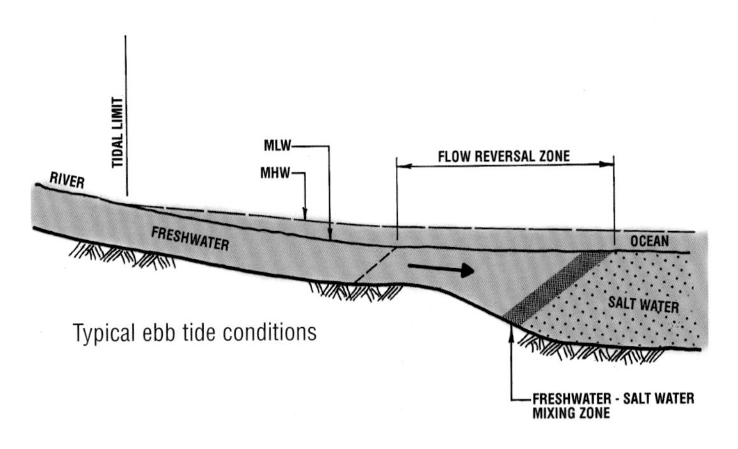

Typical ebb tide conditions

of the rivers and upstream along their channels. Because it is generally colder and denser than fresh water, salt water often advances as a stratified layer along the bottom of the channel beneath the lighter fresh water. (See Figure 37.)

Tidal influence usually extends upstream beyond the farthest point where salt water penetrates. The advancing flood tide increases the volume of water in the channel and, in effect, "blocks" the freshwater runoff from upstream, causing the freshwater segment to be deeper than it would in the absence of tidal influence. For example, the mixing zone on the Connecticut River extends up to the East Haddam bridge, 14 miles from Long Island Sound, but tidal influence is found all the way up to Enfield, some 60 miles from the Sound. During periods of high runoff, however, the elevation of the freshwater runoff is higher than the normal river level at high tide, and the vertical tidal influence is greatly reduced. Consequently, the Connecticut River's tidal cycle is noticeable during periods of low flow, but is less evident after rainstorms.

Tidal influence on a particular river is a function of many variables. Rivers with high freshwater flow rates have less tide range than rivers with low flow rates. For example, some rivers, such as Carolina Creek in East Haven, are almost all salt water. Essentially they dry up at low tide, their channels becoming little more than mud flats. Tidal rivers with little freshwater runoff often have higher flow velocities during the rising tide than during the falling tide.

There are two types of tidewater profiles (the elevation of the river surface at high tide) along coastal rivers. The high tide water elevations for rivers with small watersheds and little freshwater discharge will decline in the inland direction. This is due to the frictional resistance of the channel, which restricts the incoming tide's flow upstream and results, in effect, in the tide's "piling up" on itself.

In contrast, on large rivers, during periods of high runoff, the high tide water profile increases in the upstream direction due to the combined volume of fresh water and salt water in the channel. Such is frequently the case on the Thames, Housatonic and Connecticut rivers.

Coastal storms and tidal surges associated with easterly winds can have a major impact on tidal rivers, driving extra salt water into their channels and raising water levels. This adverse influence, which contributes to flooding, may extend inland for many miles, especially when the freshwater runoff is blocked by the tide at the same time.

Fortunately, such a combination does not occur often.

During the ebb tide, a river's discharge includes both its upland runoff and salt water that entered the channel during the flood tide. This combined discharge can create high flow velocities and cause scour.

Field measurements indicate the water elevations in tidal channels and adjacent marshes are not uniform. Even at mean high water, the water elevations in the main channels can be higher than in the mosquito ditches and on the marsh surface, where friction resistance is higher. Also, the thick vegetation along the channel banks tends to trap sediments and create small levees that help confine tidal flows in the channel, limiting the overland flow across adjacent marshes.

Tidal flow can be impeded by structures such as tide gates, undersized culverts and, less commonly, old millpond wells and dams.

A tidal river and adjacent salt marsh.

TIDAL RIVER HYDRAULICS

The evaluation of flood hazards, bridge designs and marina locations and the restoration of salt marshes can require the hydraulic analysis of tidal rivers to predict future water elevations, velocities and flow directions. Tidal river hydraulics are much more complicated than those of non-tidal rivers because of variable water elevations, flow reversals, and other unsteady conditions. In addition, some tidal systems have multiple channels that may interconnect with each other and the sea.

Freshwater discharges, tide elevations, and complex geometry generally require the use of unsteady non-uniform flow analysis with one- or two-dimensional computer models. This often demands the use of on-site tide gauges to obtain local data.

Water quality studies of deep tidal rivers and estuaries can require the use of three-dimensional computer models that evaluate the vertical and horizontal movement of water and pollutants. These computer models are used where stratified flow occurs.

The hydraulic evaluation of salt marshes and their tidal channels entails decisions regarding the movement of water through dense vegetation, probable flow paths, and ineffective flow areas. This requires, in addition to the scientific formulas, judgments that can be based only on experience.

Predictions of potential flooding made by the Federal Emergency Management Agency along tidal rivers are not necessar-

TABLE 55

FACTORS THAT REDUCE COASTAL RIVER TIDE RANGES

- Friction caused by rough streambeds and banks
- Friction due to vegetation
- River mouth sediment bars
- Channels that diverge upstream
- High upland runoff rates
- Narrow river mouths
- Artificial barriers (culverts, piers, dams)

This small tidal creek has typical tidal marsh vegetation zones along its channel, including low, high, transition, and uplands.

Large meandering tidal river with broad marsh.

ily the result of detailed studies, and actual flood elevations may be higher or lower than FEMA's figures. More precise analyses may be required to ensure the safety of new development or the adequacy of new flood-control measures.

CHANNEL MORPHOLOGY

Coastal rivers with small watersheds are often alluvial, their shape, size and sediments governed more by tidal action than by their relatively small volume of freshwater runoff. Their size is determined by scour and deposition, as affected by flow rates, velocities, sediment load, and tidal circulation. They often have a shallow, flat bottom, steep banks (due to heavy root structures in peat), and a meandering alignment. Because sediments are transported in both directions by the reversals of water flow, scour and deposition patterns can be unstable.

Rising tidewaters flow into an empty, dry channel more easily than falling tidewaters flow back to the sea. Thus, sediment can be carried into the channel more efficiently than it is carried out. The sediment settles to the bottom, especially during periods of slack water, and accumulates in the downstream segments of these small rivers.

Pine Creek in Fairfield is a good example of this, with beach sand extending 500 feet upstream of the mouth. This inland movement of beach sediments helps to explain why many coastal rivers have firm, sandy beds even though their tributaries and mosquito ditches have soft, muddy beds.

In all instances, where the inlet of a tidal river has sandy beaches on either side, there will be ebb tide and flood tide deltas for some distance above and below the inlet. Such sandy deposits are present only because of the strong tidal currents in the vicinity of the inlet. Otherwise, the current velocity in tidal rivers and coves is low. This allows for the deposition of fine-textured sediment, the dominant type being estuarine muds. These muds are about 10 percent organic material and accumulate at a slow rate (several millimeters per year). Changes in the location of this material allow scientists to track the rate of sea level rise.

Many larger tidal rivers have abundant sediment loads from upland areas as well. The coarser sediments settle within the channel. Finer sediments of silt and clay are carried farther from the channel into adjacent wetlands by high tides or are transported out to sea at low tide.

In alluvial saltwater tidal channels, the cross-sectional area of the inlet is proportioned to the tidal prism and sediment load. Large tidal prisms are associated with higher flow velocities that tend to create and sustain larger inlets, while high sediment loads from upland sources and drift of beach sand along the shore both tend to fill inlets and reduce their size. The size of the inlet is critical in limiting upstream movement of flood waters.

SALINITY

Water is classified as fresh if it has less than 0.5 parts per thousand of salt. The typical salinity of Atlantic Ocean water is about 34 parts per thousand. In contrast, that of Long Island Sound varies from 20 to 30 parts per thousand, due to dilution by freshwater outflow from rivers. Salinity is higher in the eastern Sound, near the open ocean, where there is less such dilution.

The tides bring salt water from Long Island Sound into tidal rivers where it mixes with freshwater runoff. The resulting salinity levels in tidal rivers and marshes may range from completely saline to totally fresh, depending on local conditions. These include the amount of freshwater runoff, groundwater discharges, depth, temperature, evaporation, transpiration, and tidal exchange ratios.

Salt water is denser (heavier) than fresh water and may form a separate layer beneath fresh water. In shallow tidal channels with low runoff rates, such as Carolina Creek, the salt water tends to mix well with the fresh water, resulting in uniform salinity levels from top to bottom. In deep channels, where fresh water and salt water can flow in different directions at different velocities, salinity levels can become stratified, generally increasing with greater depth.

Rivers with large watersheds will have high freshwater discharge rates after rainstorms that can push the salt water downstream beyond the river mouth. The freshwater plume of the Connecticut River, for instance, extends into Long Island Sound during high discharges. During dry weather, the saltwater mixing zone moves upstream into the channel.

In a study done in Old Saybrook, Connecticut River salinity was measured at zero parts per thousand at all depths on August 10, 1990, at high tide during a period of high runoff. Less than a month later, during a spring tide on September 7 when freshwater runoff was low, salinity was recorded at 26 parts per thousand at the surface to 31

FIGURE 38

SALINITY VERSUS DEPTH — OLD SAYBROOK, CONN.

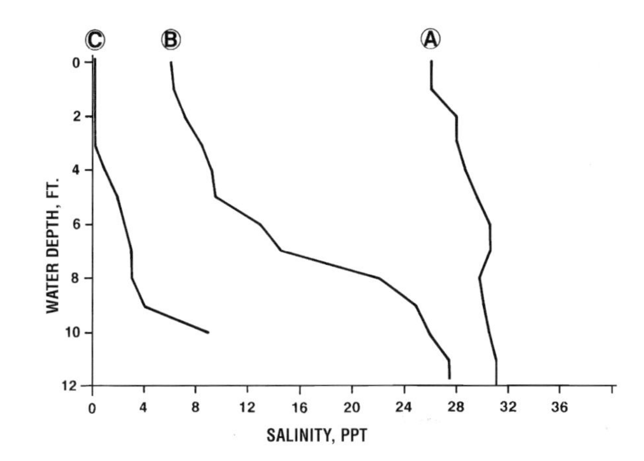

Curve A — Connecticut River at Baldwin Bridge, fairly uniform salinity curve showing a well-mixed estuary condition during high tide (9-7-90).

Curve B — Ragged Rock Creek, with stratified conditions with salinity increasing with depth (9-7-90).

Curve C — Ragged Rock Creek, with modest salinity levels due to high runoff (8-10-90).

FIG. 39

THE RELATIONSHIP BETWEEN MARSH TYPE AND AVERAGE ANNUAL SALINITY

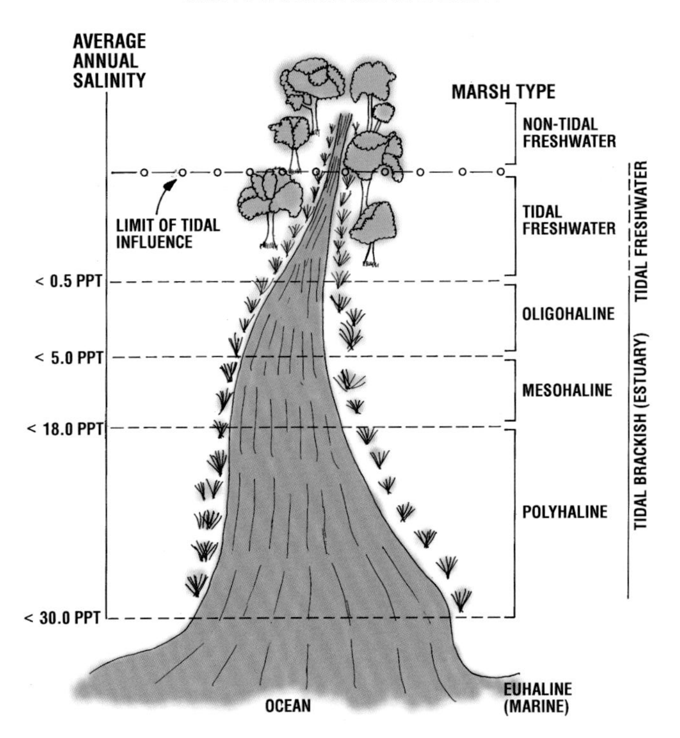

parts per thousand at a depth of 12 feet.

In nearby Ragged Rock Creek, whose much smaller watershed does not produce sufficient runoff to completely block the penetration of tidal waters, salinity was modest during the high runoff of August 10, and showed some stratification at lower depths. The September 7 conditions brought high stratification, with salinity ranging from six parts per thousand at the surface to 27 at a depth of 12 feet. (See Figure 38.)

TIDAL ZONE ECOLOGY

The one environmental factor with perhaps the greatest influence on the distribution of plants and animals in the tidal river is salinity. Three zones are commonly recognized: fresh-tidal, brackish and salt.

The salt or polyhaline zone, has a salinity range of 18-30 parts per thousand and is best known by one characteristic habitat, the salt marsh. These are the short grasses found near Long Island Sound that are dominated by four graminoid plants: saltwater cord-grass *(Spartina alterniflora)*, salt-meadow cord-grass *(Spartina patens)*, spike grass *(Distichlis spicata)*, and black grass, or rush *(Juncus gerardii)*. The last three species occur on the higher elevations of the marsh that are irregularly flooded by the tides, notably during the spring tides when the moon is new or full. Saltwater cord-grass grows at the lower elevations of the marsh along creek banks, which are flooded daily by the tides.

The high salt content also affects the types of aquatic animals found in the marshes and tidal creeks. Examples include the salt-marsh fiddler crab *(Uca pugnax)*, salt-marsh snail *(Melampus bidentatus)*, and several species of amphipods or fairy shrimp. Many of these animals feed on decaying plant material.

During flood tides, small bait fish, especially killifish, enter the marsh in large numbers and forage upon marsh snails and amphipods, thus transferring energy from the marsh to estuarine waters.

Except for the very mouth, which may be sandy, the dominant sediment in the channel of a tidal river is an estuarine mud that supports an assemblage of organisms very distinct from those in other portions of the river. In this mud are found hard clams, soft clams and various marine worms. At new and full moon during summer months, certain species of marine worms move out of the mud and into the water to spawn. Even though these worms are small, certain fish, such as the striped bass, will feed almost exclusively on them. Oysters are found on the banks of creeks and occasionally form oyster beds over the mud bottom.

Once found throughout Long Island Sound, the highly productive submerged community dominated by the flowering plant known as eelgrass *(Zostera marina)* is only found in the eastern Sound today. These dense beds support a unique assemblage of organisms, such as the pipefish, green eelgrass shrimp, and sometimes seahorses. It is critical habitat for bay scallops; the young, fingernail size scallops find refuge in the dense grass from predators. The decline of this habitat in other parts of the Sound is appartently the result of overabundant plankton growth, caused by nutrients from sewage-treatment plants, that reduced light penetration below levels required by eelgrass.

Upstream where the salinity decreases below 18 ppt is found the brackish zone. At the highest salinities, tidal marshes here also support a short-meadow-grass environment, which closely resembles that of the salt marsh. As the salinity decreases, there is an increasing abundance of tall grasses and reeds such as bulrush *(Scirpus robustus)*, three-square *(S. americanus rom pungens)*, and the most characteristic plant, narrow-leaved cattail *(Typha angustifolia)*. In many locations, the native plants are being rapidly displaced by the non-native form of common reed *(Phragmites australis)*, introduced from Europe. Two of the characteristic animals in the brackish zone are the golden ambersnail *(Succinea wilsoni)* and the red-jointed fiddler crab *(Uca minax)*.

The shallow subtidal zones (see Table 56 and Figure 40) at the higher salinity range support two submerged aquatic plants, widgeon grass *(Ruppia maritima)* and horned pondweed *(Zannichellia palustris)*. At lower salinities, the number of species increases and includes tapegrass *(Vallisneria americana)*, pondweed *(Potamogeton crispus)*, waterweed *(Elodea canadensis)*, and the non-native Eurasian milfoil (*Myriophyllum spicatum*).

In the absence of salt water occur the fresh-tidal marshes, the rarest of the tidal wetlands, which are noted for their high diversity of plant species. While these marshes superficially resemble emergent inland wetlands, the ebb and flow of the tides create an unusual hydrology and a characteristic assemblage of plants. In the low-marsh zone occurs the tall annual grass known as wild rice (*Zizania aquatica)*, the same rice that was harvested by Native Americans. In this same zone may occur the colorful, rare plant known as golden-club (*Orontium aquaticum*).

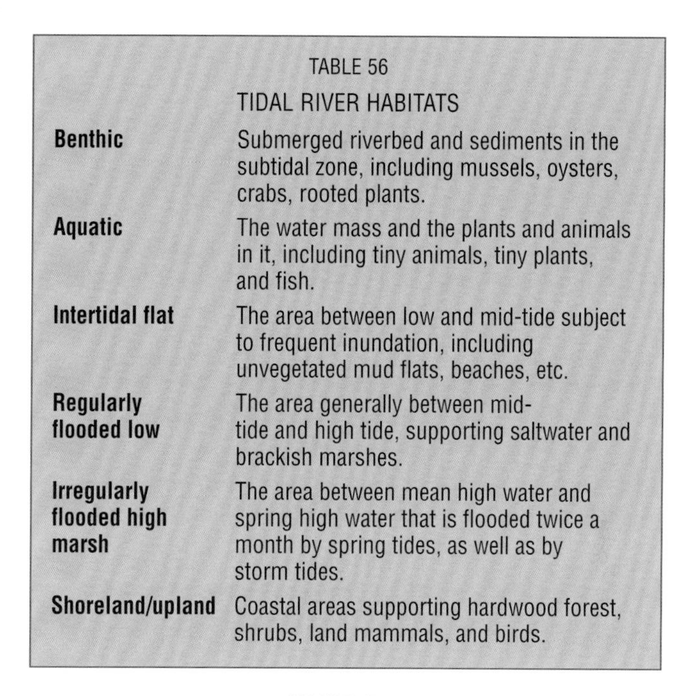

TABLE 56

TIDAL RIVER HABITATS

Benthic	Submerged riverbed and sediments in the subtidal zone, including mussels, oysters, crabs, rooted plants.
Aquatic	The water mass and the plants and animals in it, including tiny animals, tiny plants, and fish.
Intertidal flat	The area between low and mid-tide subject to frequent inundation, including unvegetated mud flats, beaches, etc.
Regularly flooded low	The area generally between mid-tide and high tide, supporting saltwater and brackish marshes.
Irregularly flooded high marsh	The area between mean high water and spring high water that is flooded twice a month by spring tides, as well as by storm tides.
Shoreland/upland	Coastal areas supporting hardwood forest, shrubs, land mammals, and birds.

FIGURE 40

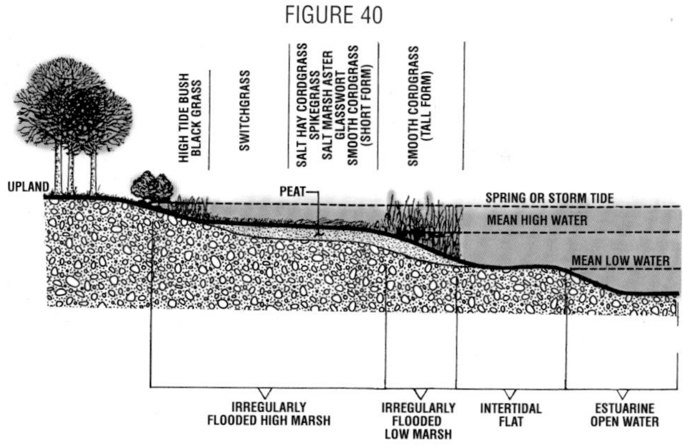

Cross section of salt-marsh zones

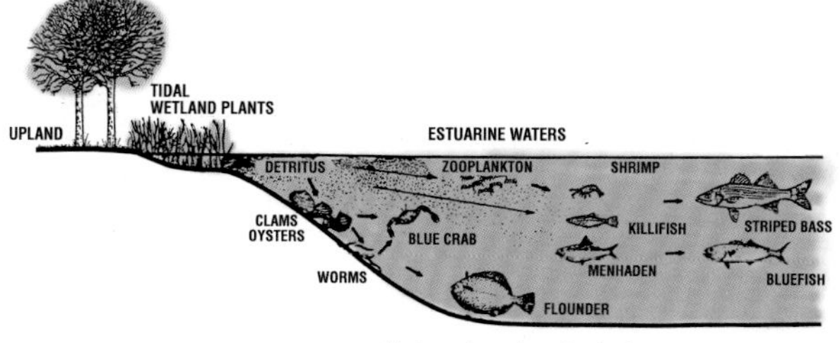

Estuarine food chain

TABLE 57

SALT MARSH FISH

FISH THAT SPEND MOST OF THEIR LIVES WITHIN THE MARSH:

Common name	Scientific name
Atlantic silverside	*Menidia menidia*
Mummichog	*Fundulus heteroclitus*
Striped killifish	*Fundulus majalis*
Sheepshead minnow	*Cyprinodon variegatus*
Four-spined stickleback	*Apeltes quadracus*
Three-spined stickleback	*Gasterosteus aculeatus*
Common eel	*Anguilla rostrata*

FISH THAT USE THE MARSH MOSTLY AS A NURSERY AREA:

Common name	Scientific name
Winter flounder	*Pseudopleuronectes americanus*
Tautog	*Tautoga onitis*
Sea bass	*Centropristes striata*
Alewife	*Alosa pseudoharengus*
Menhaden	*Brevoortia tyrannus*
Bluefish	*Pomatomus saltatrix*
Mullet	*Mugil cephalus*
Sand lance	*Ammodytes americanus*
Striped bass	*Morone saxatilis*

Source: The Ecology of Regularly-Flooded Salt Marshes of New England, United States Fish and Wildlife Service, 1986

The high-marsh meadows support a great variety of plants, including river bulrush *(Scirpus fluviatilis)*, sensitive fern *(Onoclea sensibilis)*, common sneezeweed *(Helenium autumnale)*, arrowarum *(Peltandra virginica)*, and bur-marigold *(Bidens cernua)*.

The shallow subtidal zones support submerged aquatic beds that include about 20 species. The most abundant is tapegrass. Other plants include waterweed, the non-native Eurasian milfoil *(Myriophyllum spicatum)*, and ten species of pondweed.

Although the biological communities may vary with salinity, all tidal wetlands have similar functions, such as trapping sediments, flood storage, biological productivity, energy transfer to the tidal river, and removal of nutrients from and enhancement of oxygen levels in floodwaters.

The dominant zone along the coast is the salt zone. Brackish and fresh-tidal zones are better developed on the larger rivers such as the Housatonic, Quinnipiac and Connecticut. The Connecticut River supports the best examples in the Northeast and in October 1994 was designated a wetland of international importance.

Tidal marsh soils usually are composed of peat, a mixture of sediments and partially decomposed organic material. The peat surface has black fibrous detritus and roots, underlain by older deposits of soft saturated material. The low marsh soils nearest to tidal rivers or bays may have a higher sand content, while the inner parts of a marsh have a higher silt and organic content. Both have a high water content and are usually saturated below the surface layer. The decomposition of organic material and the oxidation of iron compounds produce anaerobic conditions that limit the kinds of plants and animals that can live there. Bacteria convert sulfates carried in by sea water into sulfides, producing the characteristic marsh gases that smell like rotten eggs.

Where geological or climatic changes cause the sea level to rise, increased flooding of a marsh increases the inflow of sediments, and thus a normal tidal marsh rises with the sea level.

Barren spots can form on the high marsh surface as a result of ice, fire, grazing, excess salt, or deposits of excess thatch or mats of cordgrass carried in by high tides. The barren spots are often revegetated by invasive species such as spikegrass and the salt-tolerant glasswort.

Some tidal marshes have pannes, isolated depressions free of vegetation and filled with water. Pannes have limited inflows of water and sediment, are often underlain by very soft rotten peat, and have very

TABLE 58
TIDAL MARSH CHARACTERISTICS

CHARACTERISTICS	TIDAL FRESHWATER MARSH	TIDAL SALT MARSH
PHYSICAL		
Location	Head of estuary (above oligohaline zone)	Mid and lower estuary
Salinity	Average below 0.5 ppt	Average above 8.0 ppt and below 35 ppt
Hydrology	Riverine influence and tidal influence	Largely tidal influence
Sediments	Silt-clay, high organic content, low root and peat content	More sand, lower organic content, higher peat and root content
Sediment oxygen-reduction potential	Moderate to strongly reducing	Strongly reducing (due to reaction with sulfur in seawater)
Sediment erodibility	High (especially in the low marsh)	Generally lower
Streambank slope	Low gradient, little undercutting	Steeper gradient, more undercutting
Stream channel morphology	Low sinuosity	Moderate to high sinuosity
Dissolved oxygen (summer)	Very low	Low
Dissolved sulfur	Trace (1 ppm)	Very high (2500 ppm)
BIOLOGICAL		
Macrophytes	Freshwater species	Marine and estuarine species
Macrophyte diversity	High species diversity	Low species diversity
Macrophyte zonation	Present, but not always distinct	Pronounced
Seasonal sequence of dominant macrophytes	Pronounced	Absent or minor
Benthic algae levels	Very low (less than 1 percent of net plant community primary production)	Moderate (as high as 30 percent of net community primary production)
Phytoplankton	Similar	
Plant decomposition rates	Low marsh plants: extremely rapid. High marsh plants: moderate to slow	Moderate to slow for all plants
Nutrient cycles	Pronounced spring uptake of nitrogen-oxygen and potassium-oxygen, large autumn release of reduced compounds	Processing and release (conversion from oxidized compounds to reduced forms throughout year)
Capacity to assimilate sewage effluent	Low	Moderate

TIDAL BRACKISH MARSH

These zones are transitional between Tidal Freshwater Marshes and Tidal Salt Marsh, and will have some characteristics of each.

TABLE 59

EFFECTS OF DREDGING

- Increased flow capacity
- Increased channel depths
- Destruction of shellfish beds or smothering them with sediment
- Excessively drains tidal marshes at low tide, altering vegetation growth rates and species
- Increased high tide levels
- Increased tidal flooding
- Migration of salt water upstream
- Increased turbidity and sedimentation
- Disruption of natural circulation
- Need for disposal site for dredged materials

TABLE 60

NEGATIVE EFFECTS OF CONSTRICTING A TIDAL RIVER

- Reduced upstream salinity, causing biological degradation of salt marshes
- Reduced tidal prism and exchange rates
- Reduced channel size, due to sediment deposits and increased vegetation
- Drying and consolidation of peat deposits, leading to reduced pH levels and releases of hydrogen sulfide gases
- Damage to shellfish and fin fish spawning areas
- Altered waterfowl habitat
- Increased mosquito breeding

high salinity, the result of evaporation and occasional inflow of water.

The saltwater mixing zones of tidal rivers have abundant sources of food and shelter for small fish and serve as nurseries for juvenile fish that later migrate to the sea. (See Tables 57 and 67.) Resident fish include the common killifish and silverside. Both species feed on marsh plants and animals including worms, insects and fish fry. They, in turn, are food for larger species such as striped bass.

Among the most valuable and desirable fish, both commercially and recreationally, are anadromous fish, which spawn in freshwater rivers and whose young later migrate downstream to the ocean. These include salmon, shad, striped bass, herring, and sea run brown trout. The U.S. Fish and Wildlife Service estimates that two-thirds of commercial and recreational fish catches are dependent on estuaries.

Human Impact

Human impact on tidal rivers can be even more severe than on upland rivers, because the ecosystems of the former depend on both tidal flow cycles and salt water, and changes in tide range or salinity have a substantial impact on the biological community. The degradation of habitat can be much greater and occur much more rapidly than in other ecosystems.

In tidal wetlands where humans have disturbed the soils or reduced the salinity, the invasive common reed, *Phragmites*, often becomes established. Although eaten by muskrats, this high grass is not generally favored as a wildlife food or cover. Marshes dominated by *Phragmites* are a monoculture, noticeably devoid of insects, waterfowl and marsh furbearers. Growing to a height of 12 to 15 feet, *Phragmites* has very dense stems that physically limit human access to marshes and restrict the flow of water on the surface and through ditches, further accelerating marsh deterioration. After the growing season, the dry reeds often fuel fierce wildfires; such blazes have raced through disturbed marshes in Branford, East Haven, Stratford, and Fairfield.

Dredging

The channels of many tidal rivers have been dredged to improve navigation or control flooding. Today, dredging is carefully regulated because it can harm shellfish, degrade tidal marshes, and reduce flushing rates.

Flow Restriction

A common problem is the constriction or blockage of channels by construction of undersized culverts, and wells of mill ponds that restrict flow and limit saltwater circulation. (See Table 60.)

During the 19th century, construction of milldams blocked salmon runs, eliminating this prized fish and other migratory species from many rivers in the state. Some dams have been modified to allow for the passage of migratory fish.

Tide gates are often used to restrict the movement of tides in coastal rivers and marshes, for both flood protection and mosquito control. Most tide gates employ bulkheads with openings covered by hinged flap gates. A rising tide presses the flaps closed, preventing upstream flow. As the tide falls, the flaps open outward, releasing water from the upland areas toward the sea. This effectively drains a channel, reducing its tide range, prism and velocity. On Sybil Creek in Branford, for example, the tide range was reduced from six feet to one foot in the channel, with even less range in the marsh. (See Figure 41.) The elimination of tidal flow radically changes the water chemistry and hydrology upstream of the gates. The tide range is greatly reduced, water chemistry becomes fresh to slightly brackish, and the water table in marsh soils declines several feet. The latter results in oxidation of the organic matter in the soil and pyrite (iron sulfide) causing the soil to release sulfuric acid and the acidification of soil and creek water. Organic leachates from the soil following rainstorms can cause hypoxia or anoxia in tidal creeks. The upstream marshes become nonpoint sources of pollution.

Millponds were constructed in central and western portions of the Sound to harvest the tidal power where the tide range was the greatest. Examples include Sherwood Millpond in Westport, Gorhams Pond in Darien, and Millpond in Norwalk. In these cases, the tide gates were placed on the upstream side, allowing the tides to penetrate the river so the gates could trap tidal water. The water would return to the Sound through a sluiceway to create strong currents that would drive the waterwheel and power the mill operation. When the mills were abandoned, the sluiceways were eliminated in order to maintain an artificially high water level in the pond. This in turn reduced circulation, trapped nutrients and created warmer water conditions, often favoring the production of algae, generation of odors

FIGURE 41

SYBIL CREEK TIDE GATE IMPACT – BRANFORD
JULY 23 TO 25, 1986

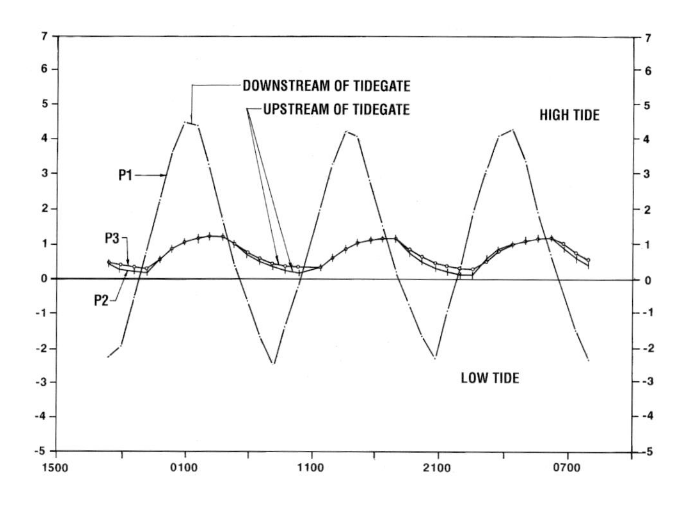

These tide gates have timbered panels hinged at the top, allowing them to swing outward to discharge water at low tide. They close at high tide due to water pressure, preventing salt water from entering the channel.

Marinas can be significant sources of pollution.

and increased sedimentation rates. The area of tidal wetland was greatly reduced to a narrow band of vegetation around the edge of the pond, a reflection of the greatly reduced tidal range.

Mosquito Ditches

By the 1940s nearly all of Connecticut's salt and brackish marshes were ditched to reduce the abundance of several species of mosquitos, especially the salt marsh mosquito. Ditches were linear and shallow with an average spacing of 100 feet. Sometimes cross-ditches were constructed perpendicular to the main ditches. Mosquitoes deposit eggs on the high marsh on moist mud that incubate for several days prior to flooding by spring tides. The larvae then develop in shallow pools called pannes. Ditches drain these pannes and lower the position of the permanent water table by several inches. In some marshes, ditches constructed in the 1930s have not been cleaned yet are still functioning today, clearly demonstrating the longevity of man-made alterations of the marsh.

While ditching did not destroy tidal marshes, it did eliminate several critical functions and values. Most notably, ditches eliminated the pannes and ponds, which represented some of the most important marsh habitat for waterfowl, shorebirds and wading birds. Also, ditches greatly reduced the stunted cord-grass vegetation of pannes, a critical habitat for the seaside sparrow, a species of special concern in Connecticut. Studies have shown that the highest diversity of bird use occurs on the natural, unditched marsh and that ditched marshes have the lowest bird use.

For these reasons, maintenance of the mosquito ditch system was discontinued in 1985 in favor of the less harmful open marsh water management techniques. The latter target only the specific areas where mosquitos breed, helping to maintain a high water table and a refuge for killifish, a fish with a voracious appetite for mosquito larvae.

Marinas

Because of their sheltered waters and proximity to the coast, tidal rivers are popular sites for marinas. However, the fuel tanks, parking lots, anti-fouling paints, repair shops, and sanitation facilities associated with marinas are potential sources of pollution.

Fixed bulkheads, piers, and breakwaters can alter circulation patterns, and dredged basins can reduce tidal prisms and exchange ratios. Dredged basins tend to accumulate fine textured sediments high in organic matter. Such sediments tend to concentrate contaminants and reduce biological diversity.

The adverse impacts of marinas can be avoided or minimized through careful siting, avoiding sensitive locations such as wetlands, minimizing depth of dredging, and reducing pollutant releases to the basin.

Stormwater Runoff and Tidal Waters

Coastal waters, including tidal rivers and marshes, can be particularly sensitive to stormwater discharges. The freshwater inflow provides nutrients to the sheltered estuaries and contributes to their high biological productivity and diversity.

The species that inhabit tidal rivers and marshes have adapted to the natural variations in flow. However, urban runoff can contribute excessive nutrients that stimulate algae blooms, and occasionally contains contaminants such as lead, copper or cadmium. Many organisms cannot adjust to the stresses of such a constantly changing environment.

The National Technical Advisory Committee has recommended that, to protect wildlife habitat, no changes in water hydrography or stormwater runoff should be allowed that permanently change an estuary's salinity patterns by more than 10 percent from natural variation. The advisory committee recommended limiting salinity variation as shown in Table 61.

TABLE 61

RECOMMENDED MAXIMUM PERMITTED VARIATION IN SALINITY OF ESTUARIES

NATURAL SALINITY PPT	VARIATION PERMITTED PPT
0.0 - 3.5	1
3.5 - 13.5	2
13.5 - 35.0	4

In developed areas, the extensive impervious surface prevents rainfall from infiltrating the ground, and storm drains convey it rapidly away. Both factors concentrate storm runoff into a shorter time period. However, base flows between rainstorms are lower than normal, since less water is retained in the soil and in wetland areas for gradual release.

Careful location of a marina, minimizing dredging, and preventing pollution releases can mitigate its harmful effects.

The effect of these unnatural extremes on coastal waters is diluted salinity levels during peak flows after storms and elevated salinity levels during the interim periods of low flow. In the case of tidal wetlands, the dilution of the salt content of the soil allows for the expansion of the invasive *Phragmites*. To minimize the impact of dilution on the estuary, stormwater projects are required to retain the first inch of runoff.

Water diversion from one drainage basin to another can have similar unnatural effects. The stream from which water is diverted by storm drains or for water supply will have reduced flow rates, causing higher

TABLE 62

BEST MANAGEMENT PRACTICES FOR STORMWATER DISCHARGES

- Pretreat runoff to remove contaminants before discharge
- Avoid extreme changes in runoff patterns and rates
- Avoid stormwater discharges that change the receiving water's salinity by more than 10 percent
- Prevent the discharge of excess sediments, toxic materials, pathogens, and floatables (solids such as plastic, and chemicals such as petroleum products)
- Disperse the runoff where possible; minimize point discharges in sensitive areas
- Do not block or obstruct tidal exchange
- Do not allow visible sediment deposits in receiving waters
- Do not block fish and wildlife movement
- Maintain groundwater recharge

salinity in its coastal mixing zone, while the basin receiving the diversion will have higher flow rates and reduced salinity in coastal waters.

The impact of excess runoff on an estuary is illustrated by the changes Hurricane Agnes caused in Chesapeake Bay in June 1972. Scientists at Virginia Polytechnic Institute who monitored the water quality and biological impact of the hurricane recorded an immediate, severe depression of salinity levels throughout the tidal system. This resulted in high mortality among populations of oysters, soft clams and hard clams, already stressed from abnormally low salinities caused by the high runoff of spring and early summer. Two heavy rains in the watershed shortly after Agnes further aggravated the problem. Altered runoff rates resulting from human activities are believed to have similar, if not such extreme, effects.

Not only does urban runoff exaggerate flow extremes, it also concentrates pollutants, which can accumulate where the runoff is discharged into sensitive coastal waters.

In Rhode Island, stormwater runoff was found to be a significant source of bacterial contamination in tidal salt ponds, with high coliform counts near developed areas after heavy rain. The Rhode Island Salt Pond Region Management Plan recommended increased use of grass swales to help filter runoff and use of holding basins where sediments and attached bacteria could settle out.

On Long Island, the U.S. Environmental Protection Agency's National Urban Runoff Program confirmed that urban runoff is the main source of high concentrations of coliform bacteria that result in closure of shellfish beds in a number of embayments (principally Great South Bay).

Since it is desirable to maintain natural stormwater runoff rates in estuaries and tidal wetlands, large stormwater discharges should not be made directly into tidal waters with limited circulation and exchange rates. Where a restricted inlet limits tidal flow (such as in the Niantic River), pollutants in the runoff could become concentrated and the fresh water could alter natural salinity.

Stormwater drainage systems at tidal marshes should distribute the runoff over a broad area and avoid point discharges. The common practice of locating storm-drain discharge points just inland of a salt marsh to avoid the need for tidal wetland permits can harm the marsh. The inland edge of tidal wetlands is usually a high marsh dominated by *Spartina patens* if undisturbed. Such areas are inundated only during the semi-monthly spring tides. Point discharges of stormwater dilute the salinity of the high marsh and encourage the invasion of brackish or upland species, notably *Phragmites*.

Drainage structures near or below the level of mean high water should be situated to avoid blocking or restricting tidal circulation. Endwalls, manholes, pipes, and fill used for access or to cover pipes should not extend into the tidal channel or above the natural marsh surface.

The use of stormwater-infiltration systems is generally desirable in coastal areas where suitable pervious soils exist. Such systems direct runoff into open basins or pervious underground structures where it can gradually infiltrate into the soil. This technique helps to reduce peak discharges in streams, recharge groundwater and improve water quality through the filtering action of the soil. Infiltration systems are used extensively on Long Island and in Florida and California to maintain higher groundwater levels, thus limiting saltwater intrusion. They should not be used in areas with on-site sewage-disposal problems or near water-supply wells.

The unique conditions found in coastal waters require the use of Best Management Practices at stormwater discharges. The principal criterion is to maintain discharge patterns and water quality as similar as possible to natural conditions, using existing discharge points or dispersing the runoff.

Restoring Tidal Rivers and Marshes

Connecticut is a national leader in the restoration of tidal wetlands, having restored more than 1,000 acres since 1980.

For salt and brackish marshes where tidal flow has been constrained by tide gates or undersized culverts, the preferred restoration approach is to reconnect the marsh to the estuary, reestablishing the natural ebb and flow of water across the wetland surface. This raises the water table to a level near the soil surface, restoring soil chemistry to a neutral or alkaline condition, arresting soil loss or subsidence, and allowing the surface to accumulate sediment to prevent drowning as sea level rises.

Typical remedial actions include removing or managing tide gates, installing larger culverts and breaching dikes.

However, such projects require careful evaluation, because simply

TABLE 63

RESTORATION TECHNIQUES

- Maintain natural circulation
- Increase tidal prism and exchange rate
- Encourage suitable tide range
- Maintain appropriate salinity in salt marshes
- Minimize upland erosion
- Protect food-chain components
- Avoid concentrated stormwater discharges
- Conserve vegetated buffer zones

removing all constrictions may lead to flooding of the soil creating a period of inundation that is too long to support vegetation.

In marshes that have subsided several feet, it is important to match the volume of tidal flow to the current soil elevations. This is done to create a flooding cycle that allows sufficient aeration of the soil to support emergent vegetation. The restoration volume is determined through hydrologic modeling and then the ideal flow regime for restoration is evaluated to determine the off-site consequences of increased flooding.

For restoration sites with structures and low-lying developed properties, engineering studies are conducted to determine the amount of tidal water required for marsh restoration. Hydraulic models are used to project the resulting tidal flood level under normal circumstances and in concert with storm tides to determine if flood-control measures are required to protect developed properties. Adequate protection can often be accomplished through a simple tide-gate management where the gates are closed during major storms and then reopened once the storms abate. Fewer situations require the construction of flood-control dikes or the use of special gates that automatically close before the tide reaches the critical flood level.

A less common restoration technique is the removal of fill or dredged sediments that were deposited on former wetlands. As the original marsh surface is uncovered and the tidal action returned, the natural restoration process begins.

In virtually all restoration projects, no planting is conducted. Rather, an abundant supply of waterborne seeds is brought in on the high tides. Over five or more years, the dense monoculture of the invasive *Phragmites* gradually succumbs to the stress of sea salts in the soil, and the native vegetation is reestablished. This restores the diversity of plants and again provides access to the marsh by wildlife, especially shorebirds, waterfowl, and wading birds. The fire hazard associated with *Phragmites*-dominated areas is also eliminated.

In 1985, Connecticut abandoned the practice of maintaining mosquito ditches. As these ditches fill with sediment, water once again ponds on the wetland surface in shallow pools that are one of the most important wetland habitats for birds. However, there is no evidence to show that the deeper, permanent ponds that were lost to ditching will return without man's assistance. Most restoration projects today include the excavation of small areas of marsh to restore this equally critical waterfowl habitat.

Restoration techniques are less established for certain brackish marshes where the disruption has resulted not from artifical alteration of the hydrology, but simply from the spread of the invasive, non-native *Phragmites*. These areas occur where high levels of freshwater runoff depress salinity to levels too low to check the spread of *Phragmites*.

RISING SEA LEVEL

Tidal rivers and coastal areas are subject not only to variations in the tide, but also to change in the level of the sea itself. The mean sea level in Long Island Sound is estimated to be rising at an average rate of 0.1 inch per year. (See Table 64.) In 1983, the U.S. Environmental Protection Agency forecast that sea level would rise at an accelerating rate, increasing as much as 2.6 feet by the year 2050 due principally to global warming. Recent studies have confirmed this trend.

A rise in mean sea level has already been recorded at gauge stations around the world. In Atlantic City, N.J., for instance, the rise was measured at 1.2 feet during the preceding 100 years, according to the EPA's 1983 report. The National Academy of Science reports that the average rise worldwide has been about 0.4 foot during the past century.

If current estimates hold true, the rising sea level can be expected to have a profound effect on coastal communities, river channels, wetlands, and drainage systems.

First and most obvious is the increase in flood levels. Present flood-level forecasts by the Federal Emergency Management Agency National Flood Insurance Program would be superseded, and existing flood-control systems would become increasingly ineffective. Waves will break closer to shore, and flood hazards from coastal storms and spring tides will extend farther inland.

Impact on Cities

Any human activities in coastal zones and along tidal rivers should take into account this projected rise in order to minimize future damages. (See Table 65.) Buildings, highways, railroads, utilities, and other structures should be placed farther from river banks and the coastline. Ten states have already incorporated predictions that the sea level will rise into their long-range planning.

TABLE 64

RISES IN MEAN SEA LEVEL 1940-1980

	FEET
Boston, MA	0.14
Wood's Hole, MA	0.36
Providence, RI	0.31
Newport, RI	0.35
New York City	0.39

TABLE 65

STRATEGIES TO PREPARE FOR A RISE IN SEA LEVEL

- Construct fewer buildings in coastal areas
- Locate all new structures on ground higher than the minimum elevation required by the Federal Emergency Management Agency
- Preserve more open space in coastal and riverine areas
- Educate the public about the forecast
- Update emergency and evacuation plans
- Design upland drainage systems to work with higher tide levels

Impact on Rivers

Higher sea levels would cause a corresponding increase in the depth of water in the downstream reaches of river channels, diminishing flow velocity and capacity to transport sediments. These changes could, in turn, cause a rise in the elevation of the streambed, reducing the slope and capacity of the channel, and increasing both the frequency and severity of flooding during periods of high runoff.

Also, higher tides would cause salt water to intrude farther upstream. This could eventually contaminate sources of drinking water, both in aquifers and where it is diverted from streams for water-supply systems.

Impact on Wetlands

Coastal wetlands, vital to shoreline communities as buffers against flooding and beach erosion, are at particular risk. These low-lying areas would be among the first to be affected by even a small rise in sea level. In eastern Long Island Sound and Fishers Island Sound, where the tide range averages 2.5 feet, the vegetation along the high marsh border appears to be changing in response to accelerated sea level rise. In particular, black grass or rush has been lost and this area is being converted to the wetter panne habitat. In the western Sound, with a tide range averaging around 7.0 feet, researchers have observed the conversion of cord-grass habitat in the low marsh to mudflat. The most extensive changes noted to date are in the brackish marshes of the Quinnipiac River, which have become noticeably wetter in the past 20 years. It is not known whether this is the result of accelerated sea level rise or other man-made alterations of the river's hydrology or sediment transport capacity.

Scientific research shows that for 2,000 to 3,000 years, tidal marshes have kept pace with sea-level rises, increasing in elevation and moving inland as sediments transported by the higher tides accumulated, building up the wetland surface.

But the extensive development along the Connecticut coast does not leave much room for further natural migration of wetlands. Fixed barriers and steep slopes also prevent marsh movement and formation.

Even in undeveloped low-lying areas, the natural sediment accumulation may not be able to keep pace with the sea-level rise fostered by global warming. If coastal wetlands are unable to follow their natural processes, prolonged submergence will cause low salt marshes to become muddy tidal flats. Present areas of high marsh will be converted to low marsh. Some upland areas will become transition tidal wetlands. In short, salt water would intrude more frequently and in greater quantities into what are now freshwater wetlands, leading to eventual dominance by plants that are salt tolerant. Interior marsh ponds will form, flooding plants and killing their roots.

On Barn Island in Stonington, exceptionally low local rates of sediment accretion are already a problem, as seen in Table 66. The present short-term rate of accretion is lower than the local rate of sea-level rise.

TABLE 66

MEASURED MARSH ACCRETION RATES AND LOCAL RELATIVE SEA-LEVEL RISE

LOCATION	SALINITY parts per thousand	MARSH ACCRETION millimeters per year RANGE	MEAN	RELATIVE RATE OF SEA LEVEL RISE millimeters per year	MEAN TIDAL RANGE meters
Barnstable, MA	20-30	3-8	5.5	0.9	2.9
Prudence Island, RI	28-32	2.8-5.8	4.3	1.9	1.1
Farm River, CT	—	—	5.0	1.9	1.8
Barn Island, CT	—	—	2.0	2.6	0.8
Great Island, CT	—	—	3.8	2.6	1.0
W. Hammock River, CT	—	—	3.6	2.6	1.4
Nells Island, CT	—	—	6.0	2.6	1.7
Stony Creek, CT	—	—	6.6	2.6	1.7
E. Hammock River, CT (Phragmites marsh)	—	—	17.1	2.6	1.4
Fresh Pond, NY	26	—	4.3	2.2	2.0
Flax Pond, NY	26	4.7-6.3	5.5	2.2	2.0

Source: National Research Council

Impact on Drainage Systems

Since higher tide levels would reduce both natural and artificial drainage, rising sea levels would threaten storm drains and sewer systems that discharge into rivers. Some gravity-flow systems would be rendered useless unless modified. Areas now above sea level might become lower than sea level and need to rely on pumping. Areas already below sea level would have to pump water farther, thereby reducing pumping capacity.

Since gravity drainage can be enhanced through the use of larger pipes or wider channels, communities should plan for this now by providing larger setbacks for roadways and structures near drainage channels.

TIDE GATES

FIGURE 42

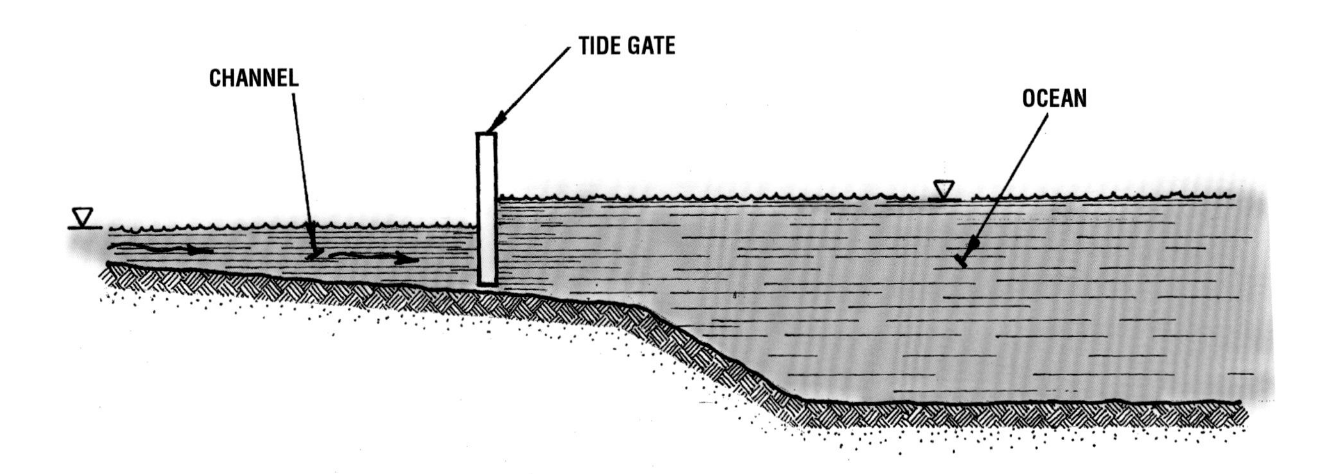

LOCAL IMPACTS

Disadvantages

- Prevent natural channel adjustments
- Prevent navigation
- Discourage mobility of fish and other sea life
- Require maintenance
- Reduce sediment transport

DOWNSTREAM IMPACTS

Disadvantages

- Raise tidewater levels near gates
- Alter sediment loads and nutrient levels
- Reduce flushing of channel

UPSTREAM IMPACTS

Advantages

- Reduce tidal flooding

Disadvantages

- Stop tidal circulation
- Lower salinity
- Can lead to salt-marsh subsidence
- Encourage brackish water vegetation, Phragmites
- Reduce habitat value and diversity
- Inhibit mosquito control

TIDE CHANNEL OBSTRUCTIONS

FIGURE 43

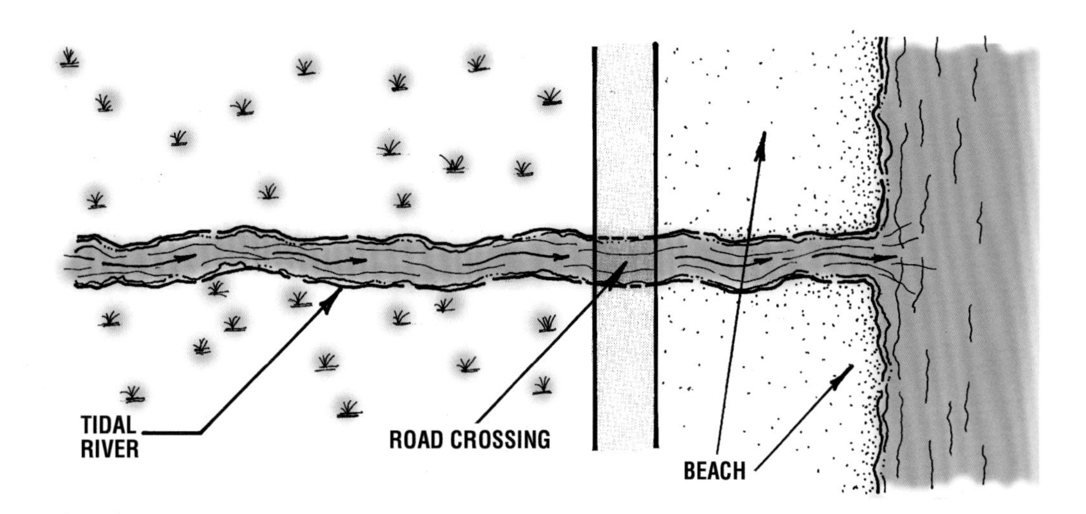

LOCAL IMPACTS

Disadvantages

- Obstruct tidal flow
- Can lead to scour
- Constrain natural channel adjustments

DOWNSTREAM IMPACTS

Disadvantages

- Concentrate tidal creek outflow
- Alter beach sediment distribution
- Encourage estuary sediment bars because of reduced tidal exchange

UPSTREAM IMPACTS

Advantages

- Reduce upstream tidal flooding

Disadvantages

- Change salt-water circulation
- Decrease salinity
- Encourage brackish-water vegetation
- Discourage salt-marsh vegetation
- Accelerate peat consolidation and settlement

TIDE CHANNEL ENLARGEMENT

FIGURE 44

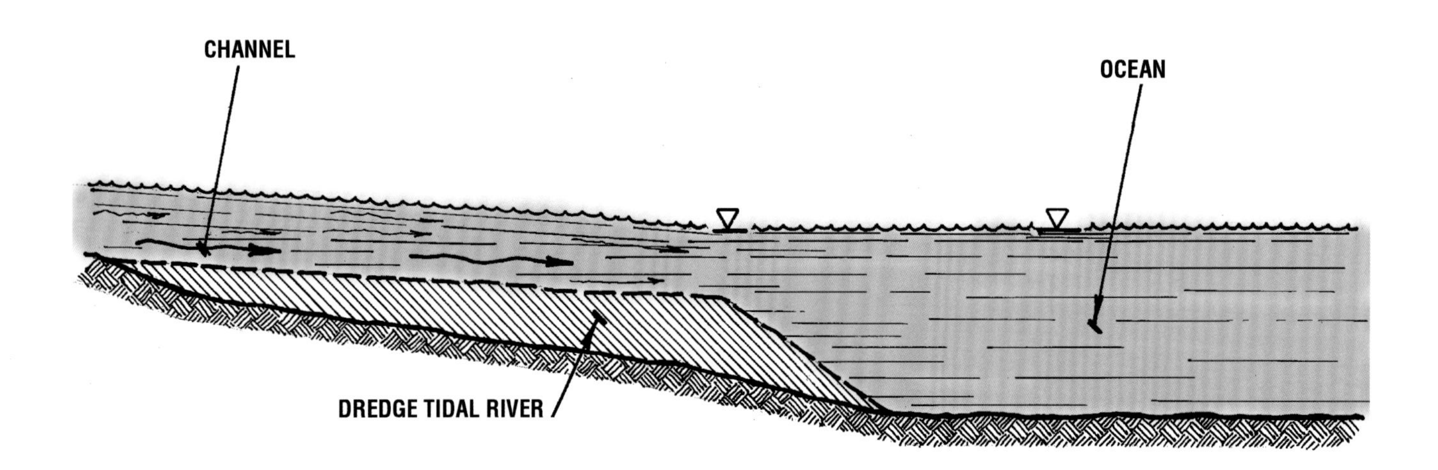

LOCAL IMPACTS

Advantages

- Increases flow depth
- Increases flow capacity
- Improves navigation

Disadvantages

- Reduces flow velocity
- Increases sediment deposition

DOWNSTREAM IMPACTS

Disadvantages

- Changes offshore beach profile

UPSTREAM IMPACTS

Advantages

- Increases salt-water circulation

Disadvantages

- Increases tidal flooding
- Changes marsh vegetation

TABLE 67

SELECTED CHARACTERISTICS OF ANADROMOUS AND SEMI-ANADROMOUS FISHES OF THE ATLANTIC COAST

FISH	SPAWNING AREA	SPAWNING TEMPERATURE °C	NURSERY GROUND	RESIDENT TIME OF JUVENILES IN TIDAL FRESH WATER	COMMERCIAL USE
ANADROMOUS					
Alewife	Small nontidal freshwater streams, also tidal fresh; primarily tributaries, on bottom	12-22.5	Tidal freshwater and oligohaline areas	Until late fall	Fertilizer
Blueback herring	In mid-Atlantic; tidal fresh and low brackish tributary (to 2 ppt)	15-24	Within five nautical miles of spawning areas	Until late fall	Fertilizer; bait fish
American shad	Nontidal & tidal freshwater in main stream; on shallow flats	12-20	Same general area as spawning area, or slightly downstream	Until late fall	Food fish, eggs for caviar
Sea lamprey	Nontidal freshwater streams in rapidly flowing water; will use tidal freshwater if passage blocked	11-24	Natal freshwater streams	3-4 years as ammocoete larvae	None
Rainbow smelt	Nontidal freshwater streams, brooks; tidal freshwater if progress blocked; main stream and tributaries	8.9-18.3	Brackish and marine waters	Little; juveniles move rapidly to sea	Food fish
Striped bass	Tidal fresh & oligohaline (to 2 ppt) in main stream; waters 2 meters deep	10-23	Same as spawning area, also associated tributaries	Until late summer	Major food fish
Atlantic salmon	Nontidal freshwater	Oct.-Dec. in New England & Canada	Nontidal freshwater	None; juveniles only migrate through tidal freshwater	Food, game fish
SEMI-ANADROMOUS					
White perch	Tidal fresh & oligohaline (to 2 ppt), tributaries and main stream shallows; also nontidal freshwater	10-20	Shallows downstream of spawning areas; mouths of tributary creeks & main stream, tidal freshwater	Slight downstream movement in summer, but may remain and over-winter in deeper channels	Food fish
Yellow perch	Mainly ideal fresh & oligohaline (to 2 ppt); less in nontidal fresh tributaries	6.8-12.5	Downstream of spawning area; lower portions and mouths of tributaries and main stream	Probably through fall	Food fish

From: "The Ecology of Tidal Freshwater Marshes of United States East Coast" by U.S. Fish and Wildlife Service, 1984.

Low tide with exposed mudflats.

High tide at same location.

Human Impact on Rivers

A river is the report card for its watershed.

Alan Levere, Connecticut Department
of Environmental Protection

TABLE 68

CHANGE IN LAND OR WATER	POSSIBLE HYDROLOGIC EFFECT
Transition from pre-urban to early-urban stage:	
Removal of trees or vegetation.	Decrease in transpiration and increase in stormflow.
Construction of scattered city-type houses and limited water and sewage facilities.	Increased sedimentation of streams.
Drilling of wells.	Some lowering of water table.
Construction of septic tanks and sanitary drains.	Some increase in soil moisture and perhaps a rise in water table. Perhaps some waterlogging of land and contamination of nearby wells or streams from overloaded sanitary drain system.
Transition from early-urban to mid-urban stage:	
Bulldozing of land for mass housing, some topsoil removed, farm ponds filled.	Accelerated erosion and stream sedimentation and aggradation. Increased flood flows. Elimination of smallest streams.
Mass construction of housing, paving of streets, building of culverts.	Decreased infiltration, resulting in increased flood flows and lowered groundwater levels. Occasional flooding at channel constrictions (culverts) on remaining small streams. Occasional overtopping or undermining of banks of artificial channels on small streams.
Discontinued use and abandonment of some shallow wells.	Rise in water table.
Diversion of nearby streams for public water supply.	Decrease in streamflow between points of diversion and disposal.
Untreated or inadequately treated sewage discharged into streams or disposal wells.	Pollution of stream or wells. Death of fish and other aquatic life. Inferior quality of water available for supply and recreation downstream.
Transition from mid-urban to late-urban stage:	
Urbanization of area completed by addition of more houses and streets and of public, commercial, and industrial buildings.	Reduced infiltration and lowered water table. Streets and gutters act as storm drains, creating higher flood peaks and lower base flow in local streams.
Larger quantities of untreated waste discharged into streams.	Increased pollution of streams and concurrent increased loss of aquatic life. Additional degradation of water available to downstream users.
Abandonment of remaining shallow wells because of pollution.	Rise in water table.
Increase in population requires new water-supply and distribution systems, construction of distant reservoirs, diverting water from upstream sources within or outside basin.	Increase in local streamflow if supply is from outside basin.
Streams restricted at least in part to artificial channels and tunnels.	Increased flood damage (higher stage for a given flow). Changes in channel geometry and sediment load. Aggradation.
Construction of sanitary drainage system and treatment plant for sewage.	Removal of additional water from the area, further reducing infiltration, recharge of aquifer.
Enlargement of storm drainage system.	Definite effect — alleviation or elimination of flooding of basements, streets, and yards, with consequent reduction in damages, particularly with respect to frequency of flooding. In stream channel, increased flood flows, lower base flows.
Drilling of deeper, large-capacity industrial wells.	Lowered water-pressure surface of artesian aquifer; perhaps some local overdrafts (withdrawal from storage) and land subsidence. Overdraft of aquifer may result in saltwater encroachment in coastal areas and in pollution or contamination by inferior or brackish waters.
Increased use of water for air conditioning.	Overloading of sewers and other drainage facilities. Possibly some recharge to water table, due to leakage of disposal lines.
Drilling of recharge wells.	Raising of groundwater level.
Waste-water reclamation and utilization.	Recharge to groundwater aquifers. More efficient use of water resources.

Modified from USGS

Human Impact on Rivers

Human activities have traditionally been concentrated near rivers because they provide a source of water, transportation, power for industry, and rich floodplain soils. Early river valley settlements became larger cities, around which suburban bedroom communities often grew up. This development and resulting changes in land use have had a dramatic impact on streams and rivers.

The effects include changes in watershed hydrology, river hydraulics, channel morphology, water quality, and ecology. Urban streams with developed watersheds differ from natural streams in that they have higher peak flows, more frequent floods, accelerated bank and bed erosion, lower water quality, and less diversity and lower populations of wildlife. (See Table 69.)

The extent and severity of the impact that human activities have on a stream are largely related to the amount and type of development within the watershed. (See Table 68.) Urban development usually includes regrading the land to accommodate roads, buildings and parking lots. Many wetlands, rivulets and intermittent streams are filled and disappear in the process. These smaller elements of the natural drainage system are replaced with artificial ones that collect and convey runoff to the larger watercourses.

Basic artificial drainage systems include gutters, catch basins and other masonry inlets, small storm-drain pipes, and swales — all designed to convey runoff from storms of relatively frequent intensity. These systems are usually close to the source areas of runoff in order to minimize drainage problems near roads, buildings and yards. They are typically designed to accommodate the peak runoff rates that occur once in every five or 10 years.

FIGURE 45

HYDROLOGIC CHANGES RESULTING FROM URBANIZATION

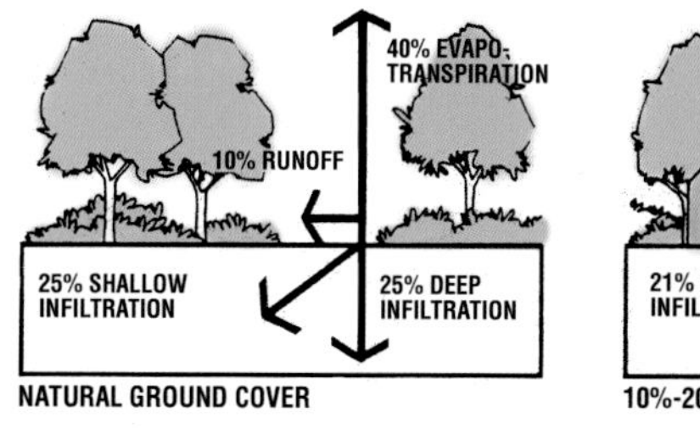

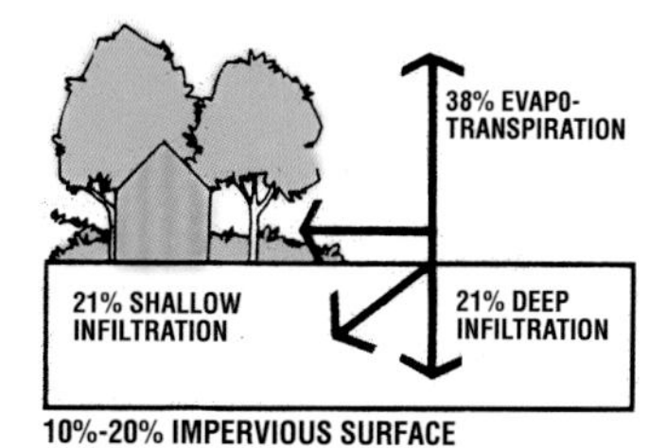

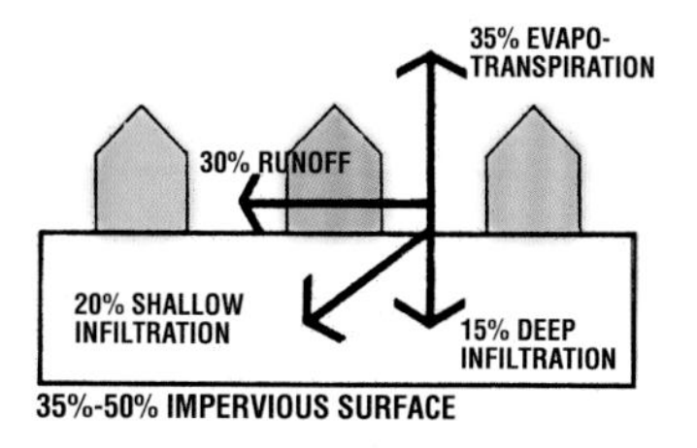

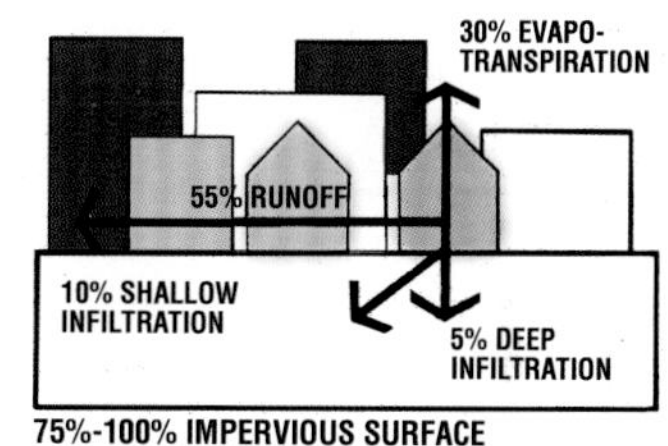

Source: North Carolina Department of Natural Resources and Community Development, in EPA, 1993.

TABLE 69

IMPACT OF LAND DEVELOPMENT ON RIVERS

Impact on hydrology

- Increases area with impervious surface
- Increases runoff volume
- Increases peak flows
- Accelerates concentration of runoff
- Reduces infiltration
- Reduces base flow
- Increases flood frequency
- Reduces floodwater storage

Impact on channel hydraulics

- Increases flow velocities
- Increases flooding
- Increases erosion
- Creates channel encroachments
- Constricts flow where bridges and culverts are installed
- Reduces floodplain conveyance
- Reduces flood storage
- Raises floodwater elevations
- Increases flood damages
- Reduces normal flow depths and rates
- Accelerates scour at existing bridges

Impact on channel morphology

- Widens channels
- Deepens channels
- Creates unstable substrates
- Increases sediment loads
- Eliminates pools and riffles
- Reduces riparian vegetation
- Increases floodplain encroachments
- Results in lined channels
- Eliminates smaller channels, concentrating flow in fewer, larger channels

Impact on water quality

- Affects nutrient content
- Raises bacteria levels
- Adds toxic materials
- Adds debris and floatables
- Contaminates bed sediments
- Increases sediment loads
- Alters water temperature
- Reduces oxygen concentrations
- Encourages eutrophication

Impact on ecology

- Reduces diversity of species
- Diminishes population of animals
- Reduces food sources
- Reduces cover for wildlife
- Reduces size and quality of aquatic habitat
- Degrades benthic habitat
- Creates fish barriers
- Causes greater fluctuations in flow
- Degrades riparian buffers
- Introduces exotic, nonnative species
- Alters riparian buffers

HYDROLOGY IMPACTS

The first land-use activities of European settlers consisted of clearing the forests for fields and pastures on small farms. Agricultural land use increases runoff by removing the natural vegetation and its resulting forest litter and porous humus soils that help retain water. Further, surface water storage is reduced by repeated plowing and smoothing of the land. Farmers also built ditches to drain wetlands and dry out their fields.

The increased industrialization and urban growth after the Civil War was followed in this century by the rapid growth of suburbs dependent on automobile transportation. Urban and suburban areas both increase the area of impervious surfaces and use artificial drainage systems to collect runoff. The prevailing philosophy for 100 years, which only recently began to change, was to convey the runoff to rivers as rapidly as possible. This reduces infiltration and evapotranspiration, increasing the volume of runoff and raising peak flow rates in the rivers. Figure 45 depicts how urbanization changes the distribution of precipitation.

In a 1968 study, the U.S. Geological Survey documented that peak runoff rates in urban areas were higher than in rural watersheds, increasing as a function of the percentage of the watershed covered with impervious surface. Most urban areas have peak flows two to three times higher than rural areas of similar size, with peak flows up to six times higher for densely developed areas with extensive storm-drain systems. (See Figure 46.)

Subsequently, the Geological Survey conducted a larger, nationwide study on the hydrology of urban flooding. Data from 269 gauged basins in 31 states confirmed that urbanization increased peak flows.

The same conclusion was reached by a study conducted in the 1980s by the Geological Survey's Connecticut office. The relationship between urbanization and peak flows was studied for six streams in four communities in Connecticut. Data were collected from July 1980 through September 1984. Seven streamflow gauges and six rain gauges were used to record data.

Using computer models based on the data, ratios of urban to rural peak flow for each area were obtained from a comparison of the simulated peak flow for storms of various recurrence intervals. The peak flows from urban areas were found to increase when the portion of the watershed with storm sewers increased. The percentage of the watershed served by storm drains is used as an indication of urban development. (See Figure 46. Note how flows increase with the increase in storm sewers and impervious surfaces.)

In addition to raising peak flows, urbanization reduces the base flows necessary for aquatic life, recreation and water supply in dry weather. A Maryland study found that the base flow declined as the impervious cover increased, with severe impacts on fish when the base flow drops to below 10 percent of the average natural discharge.

The percentage of a watershed that is covered with impervious surfaces is one of the key parameters affecting urban runoff. The greatest impact is from impervious surfaces that drain directly to watercourses and from drains that lead to watercourses. Impervious surfaces that drain to vegetated areas, such as roofs that discharge onto pervious soils, have less impact on peak flow.

The filling of wetlands during land development can also affect runoff rates. Under natural conditions, excess runoff spreads out over some wetlands. This temporarily detains the water and delays its downstream movement, reducing peak flow rates. Filling wetlands capable of storing excess water will increase peak flow rates downstream.

Although detailed hydraulic computations are required to determine the effect of wetlands on reducing downstream flow rates, several approximation methods are useful. Table 71, prepared by the U.S. Soil Conservation Service, is a simplistic assessment of how wetlands reduce peak flows.

The flood storage capacity of wetlands also is reduced by removing flow obstructions such as vegetation and forest litter. Constructing artificial channels across wetlands reduces their water retention by accelerating the post-flood drainage of pools, depressions and other ponding areas.

HYDRAULIC IMPACTS

Increased runoff into a river's channel and floodplain affect the river's hydraulics, altering its flow depth, velocities, flood frequency, scour, and sediments.

Natural, self-formed channels overflow their banks at an average interval of about once every two years. Urban streams have higher peak flow rates and therefore overflow their banks more often, increasing flood damages and erosion. A Pennsylvania study determined that watersheds that are 50 percent urbanized would have overbank flows four times more often.

TABLE 70

PREDICTED SIGNIFICANCE OF URBANIZATION ON PEAK FLOWS*

Percentage of Basin With Storm Sewers	Ratio of Urban to Rural Peak Flow for Indicated Recurrence Interval			
	2-YR.	10-YR.	50-YR.	100-YR
30	1.35	1.3	1.2	1.1
40	2.00	2.0	1.70	1.6
50	2.8	2.8	2.25	2.15
60	3.65	3.6	2.8	2.7
70	4.45	4.35	3.5	3.2
80	5.3	5.15	3.9	3.75
90	6.1	5.9	4.45	4.3

* For watersheds between 2 and 25 square miles in area

Source: USGS, 1990

FIGURE 46

EFFECT OF URBANIZATION ON MEAN ANNUAL FLOOD FOR A 1 SQUARE MILE DRAINAGE AREA

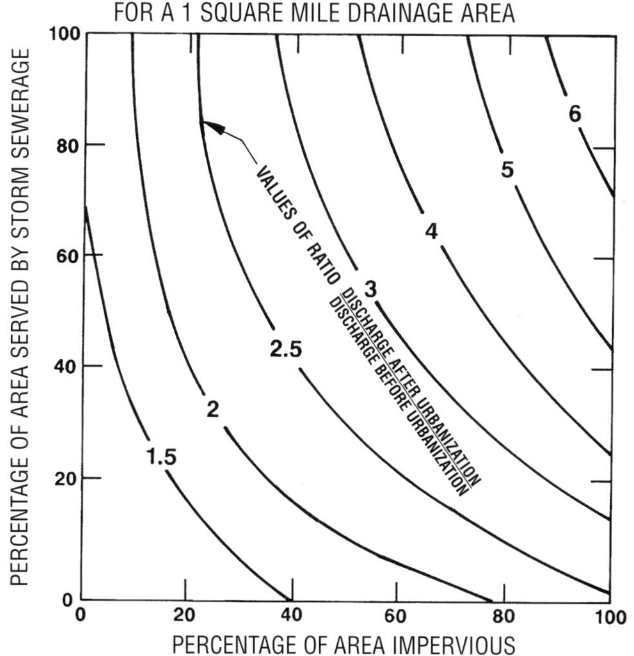

REPRODUCED FROM U.S. GEOLOGICAL SURVEY CIRCULAR 554 "HYDROLOGY FOR URBAN LAND PLANNING," 1968.

TABLE 71	
TYPICAL EFFECT OF WETLANDS ON PEAK RUNOFF	
Percentage of Watershed Covered by Wetlands	**Peak Flow Reduction Factor**
0.0	1.00
0.2	0.97
1.0	0.87
3.0	0.75
5.0	0.72

FIGURE 47

FLOODPLAIN WATER STORAGE

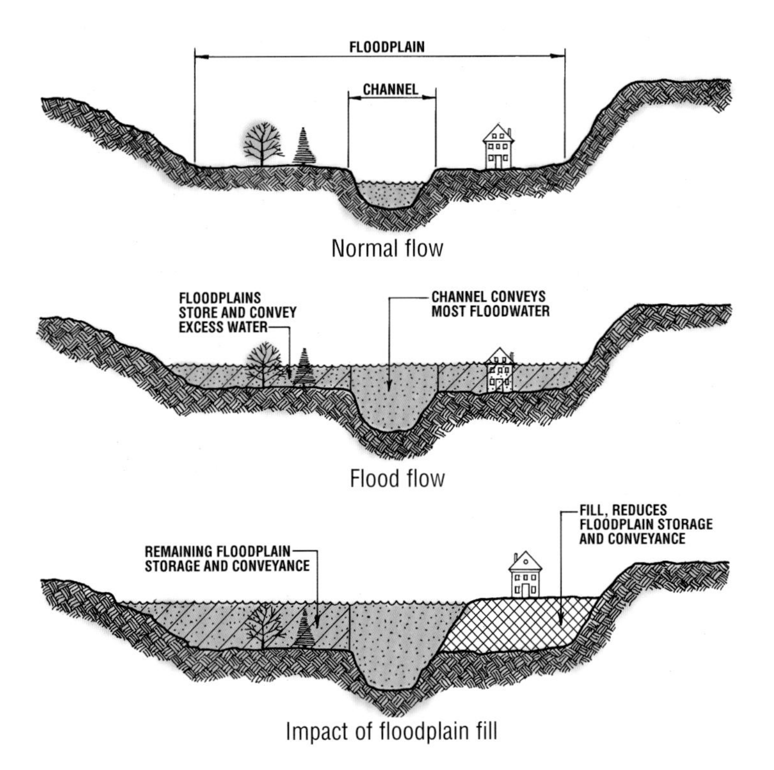

Channel and floodplain encroachments that reduce flow capacity include fill material, buildings, bridges, and culverts, among other obstacles.

Fill material is often placed along river channels to level construction sites and to gain additional land. It is also used to raise the land above floodwater levels. The hydraulic impact is a reduction in the channel's cross-section area, which reduces flow capacity. In severe cases, fill encroachments raise the elevation of floodwater upstream by obstructing its movement downstream.

Fill is also used to create dikes that confine floodwaters to the channel, reducing the movement of water onto portions of the floodplain where under natural conditions it would be stored for more gradual release downstream.

The storage capacity of floodplains is also reduced when areas subject to inundation are filled. (See Figure 47.) Although the impact of individual fill projects is usually small because of their limited size, the effects are cumulative.

Reducing the floodplain's capacity to store floodwaters can increase peak flow rates downstream, depending on the amount of fill, floodplain size, river slope, and runoff timing. Studies of the Connecticut River by the U.S. Army Corps of Engineers found that a 10 percent loss in floodplain storage would increase the downstream peak flows by 10 percent, while a 30 percent storage loss would cause a 40 percent increase in peak flow.

The loss of floodplain storage resulting from fill or dikes has a more significant effect during flash floods than during prolonged floods, because once the storage capacity is reached, the floodplain's mitigating effect is reduced. Thus it is of most benefit during brief, intense floods or during the early stages of a prolonged event.

Flood insurance and encroachment studies often employ only hydraulic analyses of steady-state flows, which do not consider the loss of floodwater storage. Nor can such methods determine the downstream impact of an upstream encroachment. Consequently, more complicated, unsteady-flow analysis methods must be used when large floodplain encroachments occur.

Road and railroad embankments that cross a floodplain perpendicular to the river can be significant barriers to flood flows. It is common practice to place bridges or culverts at the actual channel crossing, blocking the floodplain with fill material used to elevate the roadway or tracks. Such embankments can act as dams, impounding floodwaters in excess of the bridge or culvert flow capacity; if overtopped, they may be washed out or damaged as water cascades across the crest and down the other side.

However, some undersized bridges and culverts with high embankments have been found to have a beneficial effect, helping detain floodwaters and reduce the peak flow rates downstream. Studies in New Jersey have found that floodwater storage at road embankments helps to offset the increase in peak flows associated with urbanization.

IMPACT ON CHANNEL MORPHOLOGY

The hydrologic effects of land-use changes affect the shape, size and form of stream channels.

The higher flow velocities and more frequent floods scour the channel, enlarging the flow area. Urban channels have been found to be twice as large as rural channels with the same size watershed. Unless lateral erosion is contained by soil and vegetative conditions, urban rivers will generally erode their banks and increase in width. The removal of riparian vegetation further reduces the banks' resistance to erosion. Studies of 32 rivers in Maryland before and after floods confirmed that the urban rivers increased their width more than rural rivers did.

Lateral erosion leads to steeper, less stable banks that tend to be undercut and then collapse into the channel. This damages the riparian zone and makes it difficult for animals and humans to approach the water. It also adds more sediment directly into the river.

The increased channel widths carved by larger flood flows are detrimental to the river during normal flow conditions. The wider channels and reduced dry-weather flow rates result in shallow flow depths. This reduces the shelter needed by larger fish and allows more sunlight penetration to the streambed, raising water temperatures and encouraging excessive algae growth.

The higher flood flows caused by development overtop the banks more frequently and increase the flow across meander bends. Thus, alluvial streams in urban areas tend to eliminate the meanders, straightening their channels. This reduces their length and effectively increases their slope and flow velocity.

Urban streams are also known to erode their channel beds, caus-

TABLE 72

TYPICAL URBAN DRAINAGE SYSTEM COMPONENTS

Overland flow	Excess rainfall flowing across the land surface without defined channels, typically a shallow sheet of water moving through grass or over paved surfaces.
Curbs	Low barriers of concrete, bituminous concrete, or stone along edges of roads and parking lots to intercept overland flow and direct its movement.
Gutters	Shallow open channels usually associated with curbs to convey runoff along the edge of paved areas and minimize ponding on roadways.
Storm drains	Underground pipes or conduits used to transport excess surface runoff to discharge points, supplementing gutters and larger paved channels.
Drainage inlets	Masonry chambers or lateral pipes with grated openings that convey surface runoff from gutters or depressions to storm drains. Usually located at low points in yards, road intersections, along curbing, and at crosswalks.
Catch basins	Masonry drainage inlets with a sump to trap debris and sediment before it enters storm drains. Some catch basins have baffles or hoods over the outlet to the storm drain to trap floating material in the sump.
Manholes	Underground masonry structures that provide maintenance access to storm drains and serve as junctions for two or more storm drains.
Culverts	Pipes or conduits that convey runoff beneath roads or driveways from one side to the other.
Endwalls	Masonry, concrete, timber, or metal walls that reinforce the exposed ends of storm drains or culverts and help to direct water flow and minimize scour damage.
Drainage channels	Artificial linear channels constructed to convey drainage runoff, as opposed to natural watercourses. They may be lined to prevent erosion.
Swales	Natural or artificial channels that convey runoff, but with distinctive, mildly vegetated banks and bed. They usually have only intermittent flow.
Detention facilities	Natural or man-made impoundments used to temporarily store excess surface runoff and to release it gradually so that the discharge rate is lower than the peak inflow rate.

A stream that has been channelized for flood control.

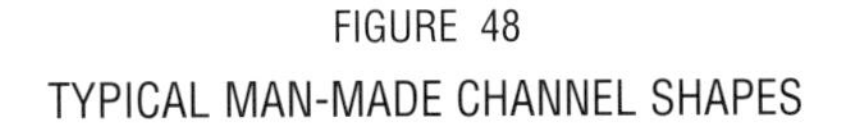

FIGURE 48
TYPICAL MAN-MADE CHANNEL SHAPES

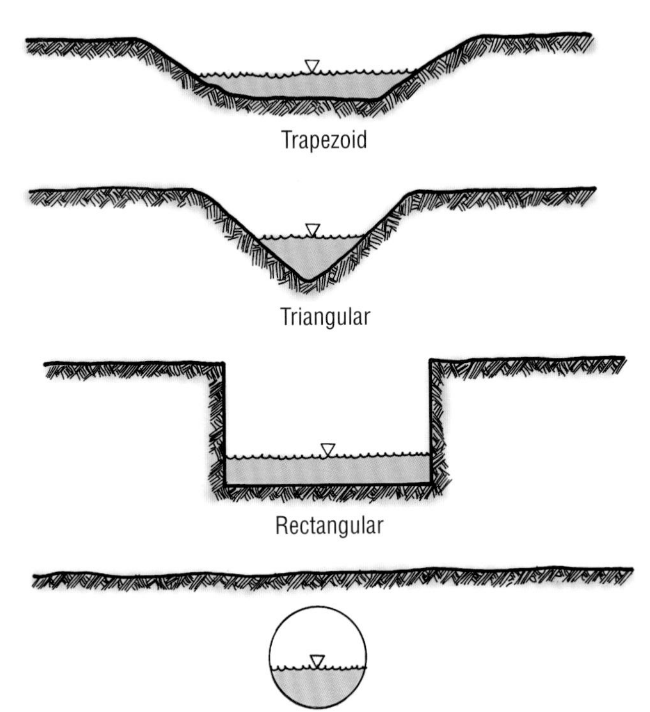

ing degradation, especially in uniform soils, at constrictions, and in channels with retaining walls along the banks.

Uniform soils may not contain enough larger cobbles to form a continuous, naturally armored bed. Consequently, bed erosion is not limited by the accumulation of native resistant material. This has been observed in the silt and clay sedimentary soils of central Connecticut.

A common reaction to urban stream erosion has been to line the channel banks with stone riprap, masonry retaining walls or steel sheeting to stabilize eroding banks or to reclaim land. This prevents channel widening and may even reduce channel width. However, channel degradation may then occur when high flow velocities erode the bed instead of the banks. This happened on the Rooster River in Bridgeport, where 2 to 4 feet of degradation occurred in just 20 years and the retaining walls were undermined. This degradation reduces the size of the riparian zone and isolates it from the river.

Channel degradation in urban areas occasionally exposes water mains, gas mains, and sanitary sewer pipes that cross under the river bed. The pipes must then be either relocated or encased in concrete to protect them from damage.

Often, bridges and culverts constructed across small streams have smaller cross-sectional areas than the channel. This restricts the flow width and increases the water velocity. As a result, bridges and culverts frequently have deep scour holes under and downstream of the structures. Many bridges have failed after their foundations were undermined by such scouring. The floods of August 1955 and June 1982 resulted in numerous bridge failures.

The material eroded from the scour holes is often deposited in the channel just downstream from the constriction as the water velocity decreases, forming sediment bars.

The depth of scour at bridges is influenced by the general bed degradation, degree of contraction, alignment, location of piers, and soil characteristics.

FLOOD CONTROL AND CHANNELIZATION

Over the years, many rivers have been dammed, diked or channelized in an effort to prevent or reduce flood damage in developed areas. (See Table 73.) Thus, the disruption that flood control projects cause in riverine ecosystems is often attributable to land use and urban development.

Several traditional flood control methods, a combination of upstream dams and downstream channelization and barriers, are employed on the Naugatuck River, which devastated communities throughout its valley during the 1955 flood. The photo on page 146 shows walls and channels in Ansonia.

Channelization is the process of widening, straightening, deepening, and otherwise altering a natural river to increase its flow capacity and reduce flood damages, as well as to improve navigation. Common practice is to construct channels with a rectangular, trapezoidal, or triangular shape and line them with concrete or broken stone. (See Figure 48 and Table 74.)

Smaller urban streams are often totally enclosed in underground pipes and conduits. (See Figure 49.) Forcing all of the water into such a smooth conduit produces unnaturally rapid movement during high-flow periods and concentrates floodwaters farther down the watershed.

Realigning rivers into straighter channels is often done to improve drainage capacity, to reclaim land, or to increase use for transportation. But straightening the channel can severely disrupt a river's morphology and ecology, eliminating natural meander bends, pools, and riffles, and greatly reducing riparian vegetation. Overall, this combination of effects destroys aquatic habitats and reduces opportunities for recreation.

The straightening of a channel between any two points reduces the length but not the vertical drop; hence, the channel slope is increased. This leads to higher flow velocities and greater sediment transport. The channel will erode and degrade unless it is artificially stabilized or an additional sediment load is supplied via upstream erosion.

Common on steepened channels, degradation can progress upstream and into tributaries. The eroded material may form sediment deposits downstream of the straightened reach.

Changing the channel slope can induce a change in channel pattern. For instance, if the bed is coarse and resists erosion, the channel will erode laterally to supply the additional sediment the river needs to achieve equilibrium. This, in turn, can lead to a braided channel that is wide, shallow and unstable. In other cases, the straightened channel may be unstable as it attempts to re-establish a meandering pattern.

Increasing the depth of a channel by dredging or excavation increases its cross-sectional area and flow capacity. However, this

TABLE 73

TRADITIONAL FLOOD CONTROL METHODS

Channelization	Increasing a river's flow capacity via greater depth and width and use of erosion-resistant linings
Storage reservoirs	Using dams and detention basins to hold excess runoff from upstream and to reduce the rate at which it flows downstream
Flood barriers	Using earthen dikes, levees, and walls to exclude water from certain areas or to confine it in channels
Diversions	Intercepting water and redirecting it along an alternate route, bypassing areas prone to flood damage

TABLE 74

CHANNELIZATION METHODS

- Enlarge width and depth
- Reduce length of channel
- Cut off meanders
- Remove vegetation and obstructions
- Provide artificial linings
- Construct dikes and floodwalls
- Enclose natural channel entirely

IMPACTS OF CHANNELIZATION

- Increased slope and flow velocity
- Reduced channel length
- Erosion during floods
- Sedimentation during low flows
- Loss of pools and riffles
- Loss of riparian wetlands, increasing downstream peak flows
- Reduced groundwater recharge, altering groundwater levels
- Loss of vegetation
- Increased water temperatures
- Removal of logs, boulders, fish shelter
- Loss of habitat diversity
- Loss of recreational use
- Reduced aesthetic value

FIGURE 49

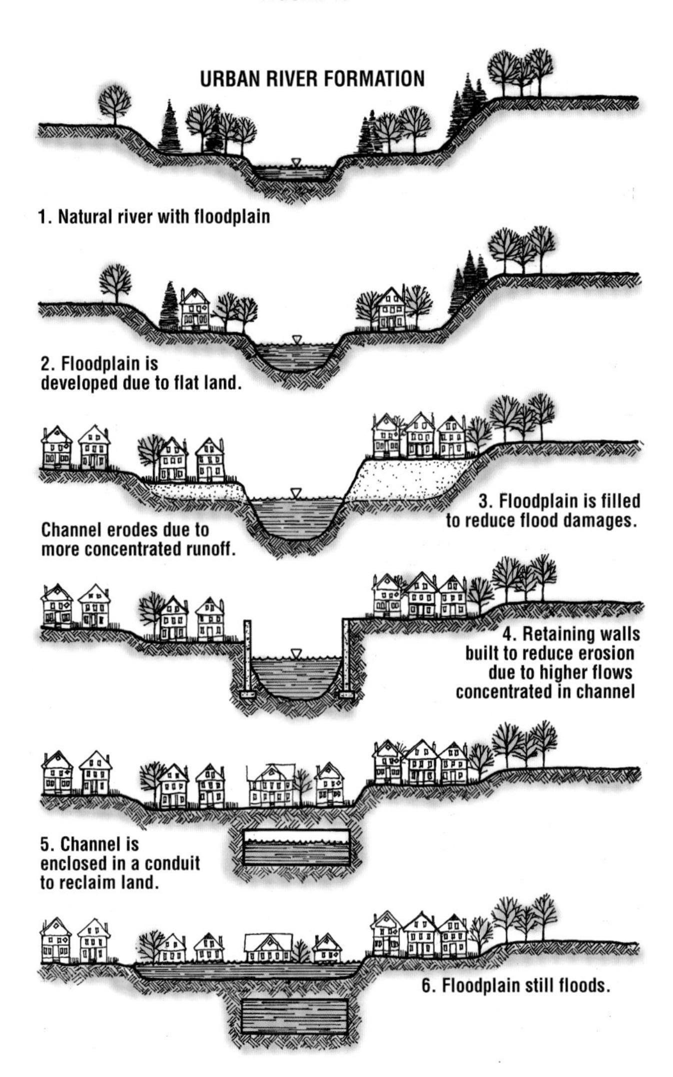

lowers the water level in the channel during normal flows, and causes a corresponding drop in the water table. This can alter subsurface water supplies and dry up adjacent wetlands that have ecological value.

Increasing the depth of the channel without changing the width will increase the slope of the channel banks. This has to be done with caution to avoid undermining the banks. In addition, the slope of tributaries will have to increase by downcutting to meet the main channel's new grade. They may then progressively scour upstream until their entire length has degraded to a new equilibrium.

The greater depth of the channel will reduce the frequency of overbank flows onto the floodplain. This reduces the flood hazards on adjacent property, but reduces temporary storage of excess water on the floodplain, which may increase peak flows downstream.

In many situations, high costs, negative environmental impact, and sometimes poor long-term performance make traditional, channelization flood-control methods inappropriate. Today, this type of flood control project tends to be limited to areas with very high flood hazards that justify stream disturbance, or to areas that have already been disrupted and have low ecological value. Alternate methods of reducing flood damages have evolved that emphasize avoiding development where flood risks are high or where it could contribute to further flooding. (See Table 75.)

Park River Flood Control

The Park River Flood Protection program, serving five towns in the Hartford area, includes many different measures to reduce flood damages. The program was developed by the U.S. Army Corps of Engineers, the U.S. Soil Conservation Service, the state of Connecticut, and the Hartford Flood Commission in response to devastating floods. Designed and constructed over a 40-year period, the program demonstrates changes in technology and approaches to flood control.

The project includes eight flood-control dams on tributary streams, channel alterations, and placing the entire main stem of the Park River and portions of its north and south branches — a total of four miles of streams — in underground conduits.

The river's main stem has literally disappeared. The first mile, from the Connecticut River to Bushnell Park in Hartford, was buried in 1944, following tremendous floods in 1936 and 1938. The river at that time was heavily polluted, an eyesore located in densely populated areas. Few people appear to have missed it.

Subsequently, the 1955 flood caused extensive damages along Park River tributaries. Additional flood-control measures were implemented, including the eight dams, located in then-rural areas where they now provide open space and recreation for residents of the suburbs that have grown up around them.

Man-made channels, usually with a trapezoidal cross-section, were constructed in suburban and medium-density areas subject to flooding. The concrete channel containing Trout Brook near Interstate 84 in West Hartford is a good example of the structural approach to flood control, without regard to environmental impact.

In contrast, flood-control measures made after 1980 to Piper Brook in Newington and to Trout Brook from Boulevard to Asylum Avenue in West Hartford demonstrate the greater emphasis placed on mitigation of environmental damage. This section of the Trout Brook project has a linear park with trails along a man-made channel that includes pools and riffles similar to those of a natural stream.

TABLE 75

ALTERNATE FLOOD CONTROL METHODS

- Regulate land use, limit impervious cover
- Preserve floodplains and wetlands
- Floodproof buildings
- Locate new buildings above flood levels
- Encourage public acquisition of flood-prone properties and maintain as open space
- Use upstream detention basins
- Encourage stormwater infiltration

ECOLOGICAL IMPACTS

Human activities affect rivers through increased flooding, erosion, fill, channelization, and discharges from point and nonpoint sources of pollution. These factors contribute to general deterioration in riverine ecosystems. (See Table 76.)

The riverbed substrate is a critical area affected by urbanization. Soil erosion from developing and urban upland areas and river channels provides an excess sediment load in urban rivers.

The potential for upland soil erosion increases as vegetation is removed during construction projects, which may increase nearby sediment loads by 10 to 100 times the natural amount. After construction is completed and the ground is stabilized, sediment levels may decrease again. However, urbanization increases the likelihood that the sediment will include contaminants, including salt, oil, grease, fertilizer, pesticides, heavy metals, and solvents.

High sediment loads upset the normal sequence of pools and riffles normally found along small streams. This eliminates both spawning areas and shelter used by aquatic life. Mayflies, caddisflies, stoneflies, and blackflies all prefer riffle habitats and will decline in numbers along

TABLE 76

ACTIVITIES HARMFUL TO RIVERINE HABITATS

- Reducing dry weather base flows
- Increasing flood flow rates
- Increasing flood frequencies
- Removing riparian and aquatic vegetation
- Reducing leaf and wood detritus
- Filling wetlands, rivulets, springs
- Increasing channel erosion
- Increasing sediment loads
- Eliminating pools and riffles
- Removing fish shelter such as logs and boulders
- Constructing dams, culverts and other barriers to fish movement
- Altering streambed substrate, eliminating spawning areas
- Installing artificial channel linings
- Dredging sand and gravel
- Reducing stream length and habitat size
- Degrading water quality
- Changing the temperature of the water

urban streams, as will populations of the fish species that depend on them. The heavier sediments carried along the channel bed of larger streams form an unstable substrate and form shifting sand bars poorly suited for the spawning and feeding areas needed by fish.

Urban rivers tend to have wide, shallow flow conditions during times of normal flow, and have fewer shade trees on the banks. This increases solar exposure and raises the water temperature. This is worsened as regrading and development destroy many of the smaller natural rivulets, streams, springs, and wetlands, thereby reducing sources of cool, oxygen-rich water.

Also lost with these small watercourses and the riparian vegetation along larger streams is the leaf and woody detritus necessary to sustain the algae and insects consumed by fish.

The reduced habitat diversity, reduced organic detritus, warmer water, and lack of shelter all contribute to a decline in the numbers of aquatic species and the populations of those that do survive.

Fish populations and diversity may decline further if barriers such as dams and culverts restrict their movement along a river to meet spawning, food and shelter needs.

Aquatic animals are not the only species harmed by development. The mammals, birds, reptiles, and other life supported by riparian vegetation must relocate when these areas are cleared, eliminating their shelter and feeding sites.

The lowered streambed and water elevation resulting from channel degradation and artificial channelization can drain surface water from adjacent wetlands and alter their habitat and wildlife. Local groundwater levels also may be reduced, harming wetland vegetation and encouraging the invasion of non-wetland species.

The reduction in floodplain flow along channelized or diked rivers can also affect the sediment composition and water quality that would normally contribute to healthier riparian and riverine ecosystems.

Floodplain flows normally have low velocities that allow some sediments to settle out in and enrich vegetated areas, rather than being transported downstream as they are in rivers disturbed by human activities. Also, overbank flows on floodplains carry many leaves and twigs to the main channel, often creating "leaf packs" used by insects at the base of the food chain. Such forest detritus is prevented from reaching the diked or channelized waterway, lessening its nutrient supply.

IMPACT OF IMPERVIOUS SURFACES

Several studies have used the percentage of the watershed covered with impervious surfaces as an indicator of changes in a stream's hydrology (flood peaks, base flow), morphology (channel size, sediment loads), hydraulics (flood frequency, water depths), water quality (temperature, contaminants), and ecology (species population, diversity).

The overall degree of stream deterioration can be influenced by any combination of those factors.

A U.S. Geological Survey study in New Jersey by Stephen Stankowski found that the ratio of impervious cover increased in relation to population densities and certain land uses. Subsequent studies confirmed that watershed runoff increased in relation to the percentage of impervious cover.

Data summarized by the Watershed Protection Institute indicate that noticeable stream deterioration occurs with a relatively low level of impervious cover in the watershed, in the range of 10 percent to 20 percent.

Research by Richard Klein of the Maryland Water Resources Administration found that the diversity of macroinvertebrates in a stream declined to "poor" levels when impervious cover reached 10 percent to 15 percent of the watershed's area, and other researchers have found poor ecological conditions with 7 percent to 10 percent impervious cover. Trout are seldom present when impervious cover exceeds 15 percent of a watershed, and the quantity of anadromous fish eggs declined when impervious cover reached 10 percent.

Table 77, based on research compiled from data by Klein and by the Metropolitan Washington Council of Governments, indicates the overall impact of impervious cover on stream quality.

Not only do impervious surfaces increase runoff volumes and rates, but research is finding that levels of nonpoint pollutants also increase with the percentage of impervious cover, and that biological diversity is reduced.

DAMS

The construction of dams is one of the more obvious human activities that can change — whether for better or worse — a river's hydrology, morphology and ecosystem.

The hydrologic benefits of large dams include the ability to regu-

Impervious surfaces increase the frequency of overbank flows, leading to greater flood damage in developed areas.

TABLE 77

IMPACT ON STREAMS FROM IMPERVIOUS COVER

% WATERSHED IMPERVIOUS	STREAM IMPACT	COMMENTS
0-10	Minimal	Limit for protecting sensitive native trout streams.
10-15	Low	Limit for protecting average streams. Degraded habitat.
15-25	Medium	Limit for controlling specific nutrients and toxic pollutants.
25-50	High	Reduced low flows, higher peak flows. Few fish.
50	Severe	Severe changes in hydrology, hydraulics, morphology, water quality. The stream will have few natural attributes.

TABLE 78

DIRECT AND INDIRECT EFFECTS OF CHANNELIZATION ON WILDLIFE

KEY

● Potential direct effect

○ Potential indirect effect

	REPTILES AND AMPHIBIANS	NONGAME BIRDS	GAME BIRDS	SMALL MAMMALS	FUR BEARERS	LARGE MAMMALS
Loss of wetlands, pools, meanders, and sheetwater	●	○	●	○	●	○
Reduction of vegetation diversity	●	●	●	●	●	●
Elimination of snags, overhangs, and debris	●	○	●	○	●	○
Reduction of aquatic vegetation	●	○	○		○	
Reduction of aquatic invertebrates	●	○	○			
Land use changes	○	○	○	○	○	○
Berm composition and slope	●	○	○	○	○	
Reduction of woody plants	○	●	●	○	○	○
Lowered water levels	●				●	
Lowered density, abundance of other wildlife	○	○	○	○	○	○

Source: "Manual of Stream Channelization Impacts on Fish and Wildlife," United States Department of the Interior, Fish and Wildlife Service, 1982.

late downstream flows by using adjustable gates or valves. Connecticut has many dams that impound stormwater runoff to reduce downstream peak flows and flood damages. The Park River basin and the Naugatuck River Valley both benefit from use of coordinated flood-control dams.

Some dams are used to store water during wet seasons and release it slowly during the summer. This helps to maintain streamflow for wastewater assimilation, fishing, canoeing, and hydropower. The Farmington River benefits from such releases from the Colebrook and West Branch reservoirs, both on the river's west branch.

Other dams create lakes and reservoirs used for drinking-water supplies or to meet industrial needs. Cedar Lake, Hitchcock Lake, Chestnut Hill Reservoir, and Woodtick Reservoir in Wolcott were all formed by dams constructed by Scovill Manufacturing Co. along the Mad River and its tributaries to supply water for the brass industry; today these bodies of water are used for recreation.

Two of the largest dams in the state form Lake Zoar and Lake Lillinonah on the Housatonic River. These huge dams were built for hydroelectric power generation. Their lakes are also used extensively for recreational boating, swimming, and fishing, and have become popular residential areas.

The adverse impacts of dams vary depending on individual site conditions. Negative effects include upstream aggradation and flooding, downstream degradation, and impeding the movement of fish.

Upstream aggradation can occur when river sediments settle as they enter the deep, still water of the impoundment. The sediment deposits accumulate and form a delta that extends upstream into the river channel. At Woodtick Reservoir, the delta has reduced the reservoir size but has become a rich wetland.

At dams with small impoundments, the sediment deposits may completely fill the pond, displacing all of the water. This has occurred at many sites in Connecticut, especially at low, run-of-the-river dams with large watersheds and small impoundments used for hydroelectric power. Some of these ponds evolve into mudflats with shallow submergence, or into vegetated wetlands. Others are dredged to remove sediments and restore the pond.

While most dams reduce downstream peak flow rates, they also raise upstream water elevations. This is a concern where development that has occurred beside the lake or stream above the dam can be damaged by flooding if the dam's spillways have limited capacity and are unable to drain off flood flows fast enough.

Because sediments settle in the still waters of the impoundment, the channel below the dam will receive less sediment than the river can carry. As a result, the downstream reach may scour and degrade until the river again achieves equilibrium by reducing its slope and its sediment transport capacity. Such degradation may lead to steep, unstable banks, a change in river pattern, and even the undermining of the dam.

For example, over the course of 90 years, the Mattabesset River in Kensington degraded more than six feet just downstream from the Kenmere Dam, despite the presence of several six-foot-deep, concrete erosion-control sills — barriers constructed across the channel to retard erosion. The combination of a reduced sediment load from upstream and a uniform soil incapable of forming an armored channel led to degradation at a rate of almost one foot per decade. The channel bed degraded so drastically that the erosion sills were destroyed, as was the masonry stilling basin apron built to absorb the energy of flowing water at the toe of the spillway.

An excellent example of a dam's effect on river morphology is the Leesville Dam on the Salmon River in East Haddam. The Salmon, which has a 100-square-mile watershed, is a steep, fast-flowing river with a sediment load that is high in bedload (sands and gravel). Several different dams have been built at the same location since the 18th century. The present dam is 140 feet wide.

Over the decades, the bedload was trapped in the dam's impoundment and by 1938 had reached an elevation near the crest of the spillway. Portions of the dam failed during the hurricane of that year and much of the sediment was washed out, completely filling the downstream channel, which had to be dredged.

Studies in the 1970s revealed that sediment once again filled the impoundment and that aggradation extended several miles upstream. During periods of low flow in the summer, the sediments are exposed and the river cuts a meandering channel through them to the dam. The channel downstream has degraded to an armored bed of gravel, which provides excellent fish habitat. However, aquatic life is periodically disturbed when storms cause high sediment discharges over the spillway.

Dams are barriers to navigation, recreational boating and canoe-

TABLE 79

CONSEQUENCES OF CHANNELIZATION ON AQUATIC SYSTEMS

	CIRCULATION AND RESPIRATION	EXCRETION, OSMOREGULATION, IONIC REGULATION	REPRODUCTION	TOLERANCE TO LETHAL OR STRESSFUL CONDITIONS	DEVELOPMENT AND GROWTH	FOOD AND FEEDING	NICHE	BEHAVIOR	DENSITY	DISTRIBUTION	PRODUCTIVITY	SPECIES DIVERSITY RICHNESS & REDUNDANCY	BIOMASS	ENERGY AND MATERIALS TRANSFER
Loss of specific substrate	○		●		○	●	●	●	●	●	●	●	●	
Removal of snags, root masses, etc.	○		●		○	○	●	●	●	●	●	●	●	
Loss of instream vegetation		●	○	●	●	●	●	●	●	●	●	●	●	
Disruption of run-riffle pool sequence	●		●		●	●		●	●	●	●	●	●	
Loss of overall stream length	○		○			●	●	●	●	●	●	●	●	
Increased gradient and velocity	○			○	○	●	●	●	●	●	●	●	●	
Dewatering of adjacent lands			○			●	●		●		●	●	●	
Change in basic physicochemical regime	●	●	●	●				●	○		●	○	○	
Decreased allochthonous input			○			●	●		●		●	●	●	●

Source: "Manual of Stream Channelization Impacts on Fish and Wildlife," United States Department of the Interior, Fish and Wildlife Service, 1982.

Artificial flood-control measures include dumped rock riprap (right), placed cut stone pavement (left), and concrete flood walls (rear).

Development of a floodplain is eventually subject to almost inevitable flooding.

ing, and to the movement of fish. They often form the upstream limit of anadromous fish migration, forcing these species to spawn in less desirable areas. Although some dams are now equipped with fish ladders to permit passage upstream, they are blamed for destroying salmon migrations in Connecticut and for limiting shad reproduction.

Dams also affect water quality in terms of thermal conditions, oxygen and nutrient levels, and algae growth.

Deep impoundments become stratified, with warmer surface water above the denser cold water. Water released downstream may be warm or cold, depending on the elevation of the intakes. Warm water can harm aquatic life, both as a result of the temperature and because it carries less dissolved oxygen. Cool water is beneficial for cold-water fish such as trout, but can be rich in nutrients from bottom sediments, and thus may stimulate excessive algae growth downstream. The use of multilevel intakes allows the outflow's temperature to be adjusted.

The concentration of nutrients in impoundments can stimulate the growth there of floating and rooted algae, which is detrimental to fisheries, swimming, and water quality when growth is excessive.

FLOW DIVERSIONS

Water diversions (defined by the Connecticut General Statutes to be any change in a stream's instantaneous flow rate) encompass a broad range of activities, ranging from the effect of dams and detention basins to consumptive uses and interbasin transfers of water.

Flow diversions can have a tremendous effect on rivers and their habitats, but they are often the only way to obtain large quantities of potable water for human use.

The principal concern regarding diversions centers on those that diminish the flow rate during dry weather, when water is needed for activities such as drinking-water supplies, irrigation, industry, waste assimilation, fisheries, and recreation. (See Table 80.)

Water allocations and diversions are among the most important resource issues in Connecticut, and their regulation is becoming more important as the quality of river water improves and demand for it increases.

In watersheds where there are intense, competing demands for water — as, for example, in the Quinnipiac and Farmington River basins — diversions must be carefully controlled to allocate the limited resources. Instream flow studies may be needed to determine the demands on a river's resources and the minimum flow required to meet those uses without degrading water quality or disrupting the stream's ecology. A typical diversion permit may require that a minimum base flow be maintained, or may limit water withdrawals to certain months or flow heights. Water conservation is being emphasized to minimize the use of this precious resource and reduce the demand for diversions.

Intrabasin diversions that withdraw and rapidly return water to a river have only a limited, local impact. A typical example is drawing water from the river for industrial or power-plant cooling and then returning it to the river at or near the point of withdrawal.

Consumptive water use within a drainage basin and diversions from one watershed to another do not return water back to the river at or near the point of withdrawal and must be carefully regulated to ensure that sufficient water remains in the donor stream to meet its needs.

Consumptive diversions include irrigation, drinking-water supply, and some industrial processes. The impact of these diversions can be lessened by limiting them to periods of excess flow. In addition, water-supply reservoirs with excess water available can be used to augment the natural low flow rates of summer.

The water needs of large cities are so great that river flows are diverted from one watershed to another to replenish their reservoirs. Connecticut's five largest cities all have large diversions to augment insufficient local sources of water. These facilities include impressive rock tunnels that convey water by gravity flow beneath watershed divides.

For instance, the South Central Connecticut Regional Water Authority, which serves New Haven, has tunnels capable of diverting water from the West River and Farm River to Lake Gaillard Reservoir, built in 1933, and a two-mile-long tunnel from Wepawaug Reservoir to Maltby Lakes.

The Hartford Metropolitan District obtains most of its supply from the huge Barkhamsted and Nepaug reservoirs, diverting their waters from the Farmington River basin to the West Hartford treatment plants via tunnels constructed under Talcott Mountain in 1914 and 1964.

The Bridgeport Hydraulic Co. acquired land and built reservoirs to meet the water-supply demands of the war years in the first half of the 20th century. The monumental Saugatuck Reservoir in Easton and Weston, finished in 1942, is connected to the Aspetuck Reservoir in Easton by a

TABLE 80

THE IMPACT OF LOW FLOW

- Reduces dilution of sewage treatment plant effluent
- Reduces recharge of groundwater aquifers
- Reduces flow depth
- Increases water temperature
- Reduces cover for fish
- Allows salt water to migrate upstream in coastal areas
- Inhibits navigation
- Reduces water available for industrial cooling
- Degrades recreational use
- Impedes fish migration
- Reduces sediment transport

1.5-mile-long rock tunnel. Water from the Bridgeport system is used in the broad area from the Housatonic River Valley to Stamford.

Waterbury also diverts large amounts of water to serve the city and suburban areas. The Pitch and Morris Reservoirs collect runoff in Morris and are supplemented by flow diverted from the Shepaug River watershed via a 7.5-mile-long tunnel completed in 1928. This massive tunnel passes deep under Bantam Lake en route to the Morris Reservoir system.

As the competition for limited resources increases, it will not be possible to satisfy all demands among conflicting uses. Management plans are needed to determine the allocation of riverine resources.

Figures 50-70 provide specific examples of the impact of human activities on the river systems.

WATERSHED DEVELOPMENT

FIGURE 50

LOCAL

Advantages:

- Economic development
- Employment opportunities
- Services and housing

Disadvantages:

- Increased impervious ground cover
- Decreased rainfall infiltration into soil
- Displacement of surface water by filling of depressions, wetlands
- Increased surface runoff volume
- Increased runoff velocity
- Possible decrease in floodplain conveyance
- Decreased base flows

UPSTREAM

Disadvantages:

- Possible rise in floodwater levels

DOWNSTREAM

Disadvantages:

- Increased sediment load during construction
- Increased runoff rates, peak flows
- Increased channel flow velocities
- Increased channel erosion after construction
- Decline in water quality due to urban runoff
- Decrease in base flow

COMMENTS

Land development is the principal human activity that alters the natural hydrologic cycle, affecting rivers, lakes, wetlands, and groundwater.

CHANNEL CLEARING

FIGURE 51

VEGETATED CHANNEL

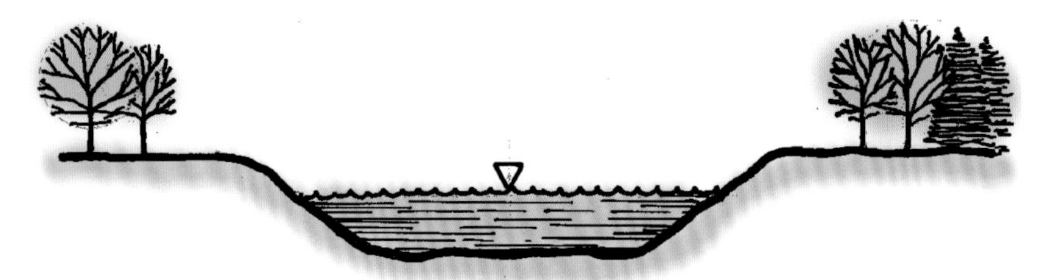

CLEARED CHANNEL

LOCAL

Advantages:

- Increased flow capacity for storms of frequent recurrence intervals
- Reduced floodwater depths
- Reduced flood damage
- Removal of excess debris

Disadvantages:

- Increased flow velocity
- Increased erosion potential
- Increased water temperature
- Decreased water turbulence and aeration
- Disturbed habitat
- Reduced instream cover and shelter for fish

UPSTREAM

Advantages:

- Lower floodwater depths

DOWNSTREAM

Disadvantages:

- Increased sediment load and deposition
- Increased water temperature
- Decreased dissolved oxygen levels
- Reduced input of leaf and woody organic matter

COMMENTS

Excess channel clearing eliminates the natural shelter and organic detritus that contribute to the aquatic habitat. The natural riparian vegetation and channel irregularities should be preserved.

CHANNEL FILLING

FIGURE 52

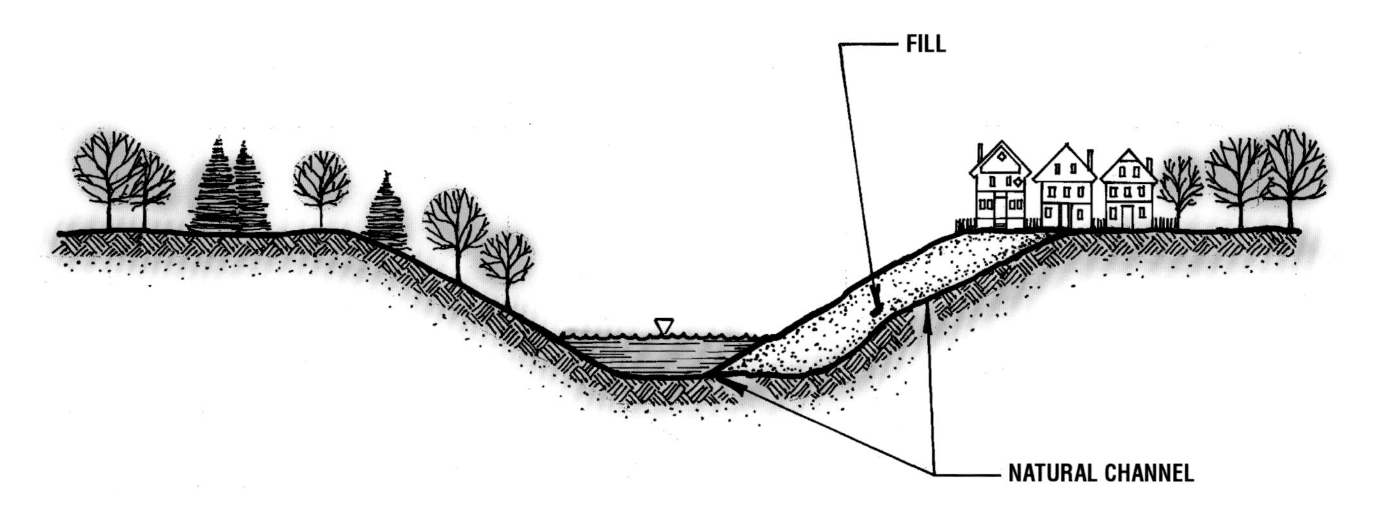

LOCAL

Advantages:

- Increased land area for human uses

Disadvantages:

- Reduced channel size
- Increased flow velocities
- Increased water depth
- Increased bed and bank scour
- Destruction of riparian habitat and elimination of shade
- Decreased floodwater storage and conveyance
- Alteration of lateral drainage of runoff to river
- Prevention of natural channel adjustments

UPSTREAM

Disadvantages:

- Increased water depth and flooding
- Possible reduction in flow velocities

DOWNSTREAM

Advantages:

- Decreased peak flow rates

Disadvantages:

- Increased sediment load leading to reduced water quality and possible aggradation
- Concentrated flow causes scour at end of fill area

COMMENTS

Placing fill material in river channels often causes an increase in floodwater levels.

CHANNEL WIDENING

FIGURE 53

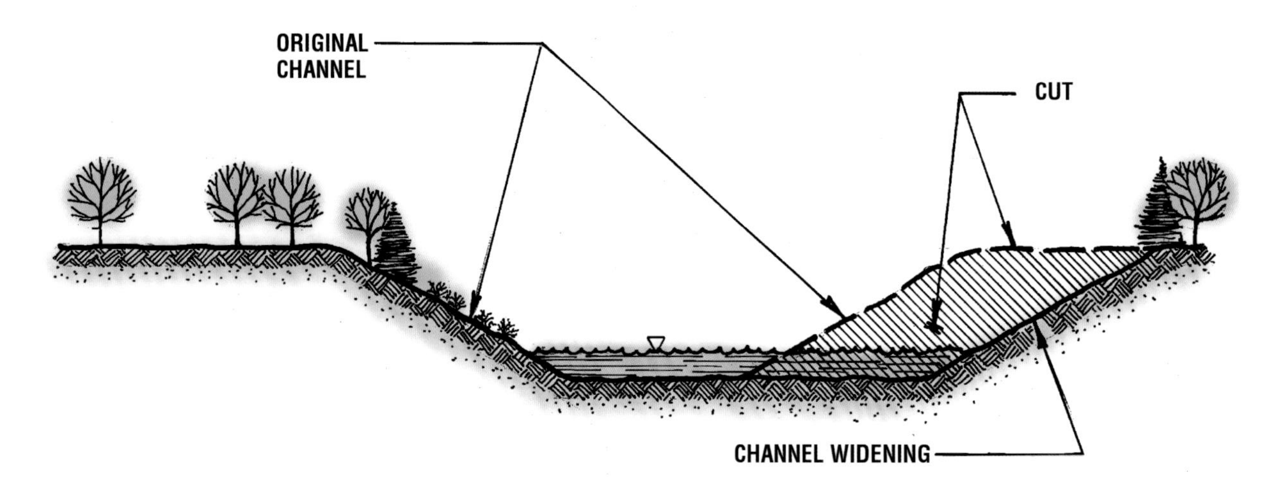

LOCAL

Advantages:

- Increased flood flow capacity
- Decreased flood flow depth
- Increased channel storage

Disadvantages:

- Decreased flow velocities
- Increased sediment deposition
- Destruction of riparian habitat
- Increased water temperature
- Potentially unstable banks
- Reduced flow depth

COMMENTS

Try to save vegetation on one or both banks.

UPSTREAM

Advantages:

- Reduced flood flow depths

Disadvantages:

- Possible channel aggradation
- Scour at transition

DOWNSTREAM

Advantages:

- Could decrease peak flows by increasing storage volumes

Disadvantages:

- Could increase peak flows
- Increased sediment load during excavation
- May cause channel degradation

CHANNEL STRAIGHTENING

FIGURE 54

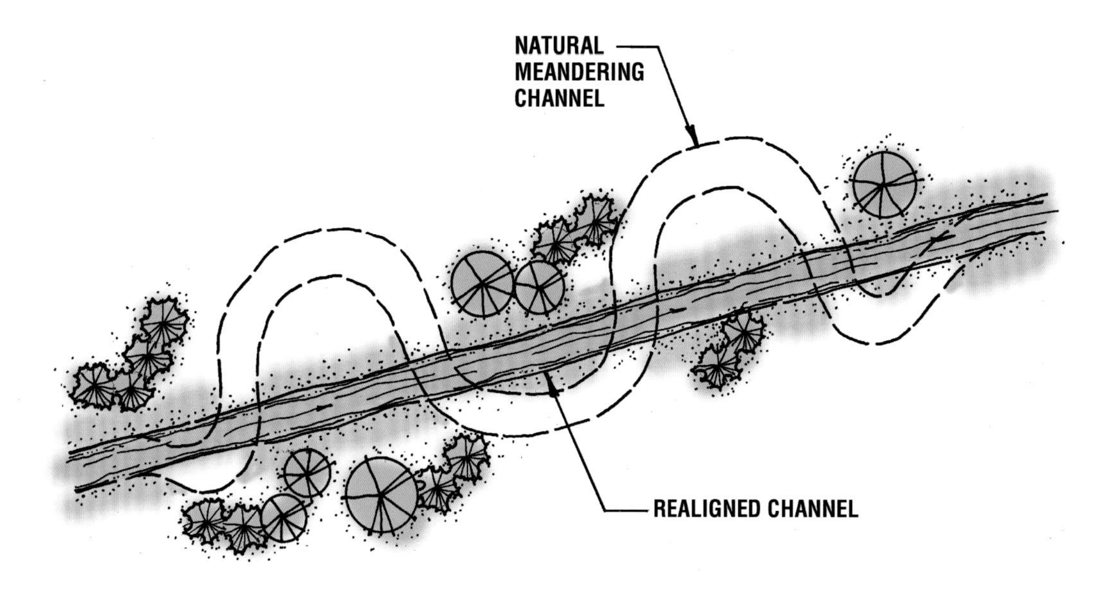

LOCAL

Advantages:

- Increased conveyance capacity
- Reclamation of land
- Possible reduction of flood damage

Disadvantages:

- Shorter channel length, increasing slope
- Higher flow velocities
- Increased scour
- Reduced floodplain storage
- Elimination of pools and riffles
- Reduced bank habitat area
- Reduced aquatic habitat area
- Channel scour protection required

UPSTREAM

Advantages:

- Lower flood stages

Disadvantages:

- Potential channel degradation

DOWNSTREAM

Disadvantages:

- Increased flood flow rates and stages
- Increased sediment loads
- Increased sediment deposition; if excavated area is not armored, may bury bed habitat

COMMENTS

The realignment and relocation of channels is usually associated with land development and highway projects.

CHANNEL RELOCATION

FIGURE 55

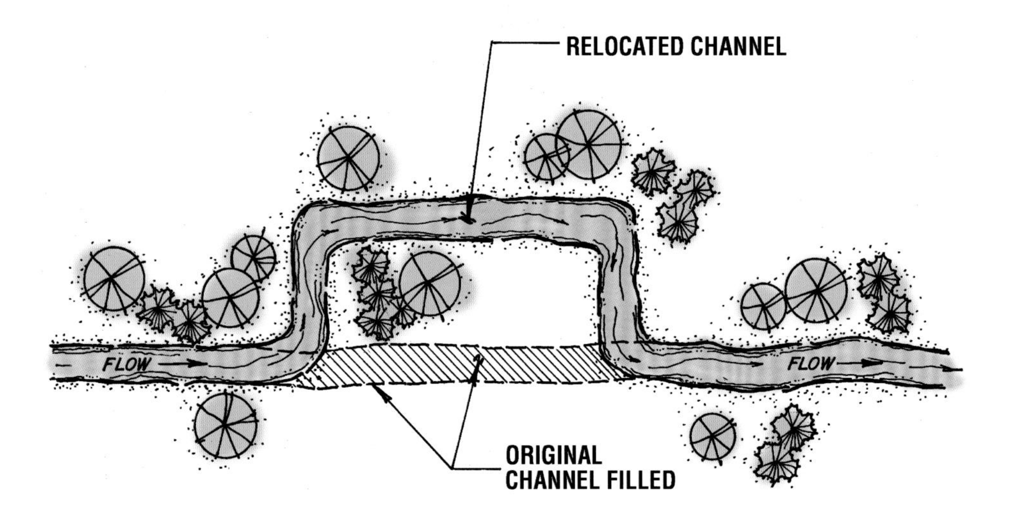

LOCAL

Advantages:

- Old channel reclaimed for human uses
- Increased flood storage if old channel not reclaimed

Disadvantages:

- Increased channel length
- Decreased channel slope
- Decreased flow velocity
- Decreased flow capacity
- Aggradation due to increased sediment deposition
- Destruction of channel habitat

COMMENTS

Careful design and creation of natural conditions in new channel can mitigate negative impacts.

UPSTREAM

Disadvantages:

- Aggradation
- Scour at transition
- Possible increase in water elevation
- Possible change in channel pattern, alignment

DOWNSTREAM

Advantages:

- Delay of peak flood flows

Disadvantages:

- Decreased sediment load, leading to degradation

CHANNEL REALIGNMENT

FIGURE 56

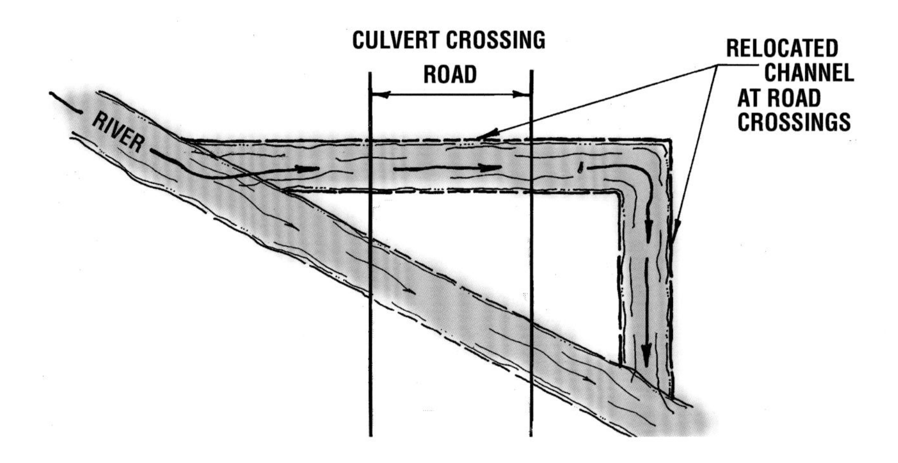

LOCAL

Advantages:
- Allows culvert to be shorter
- Increased channel length, less slope
- Lower flow velocity

Disadvantages:
- Potential erosion in new channel
- Unnatural, abrupt bends, prone to erosion
- Reduced natural habitat
- Channel linings often required

UPSTREAM

Advantages:
- Reduced flow velocity

Disadvantages:
- Higher water elevations

DOWNSTREAM

Disadvantages:
- Increased sediment load

COMMENTS

The natural channel alignment should be maintained wherever possible.

CHANNEL DEEPENING

FIGURE 57

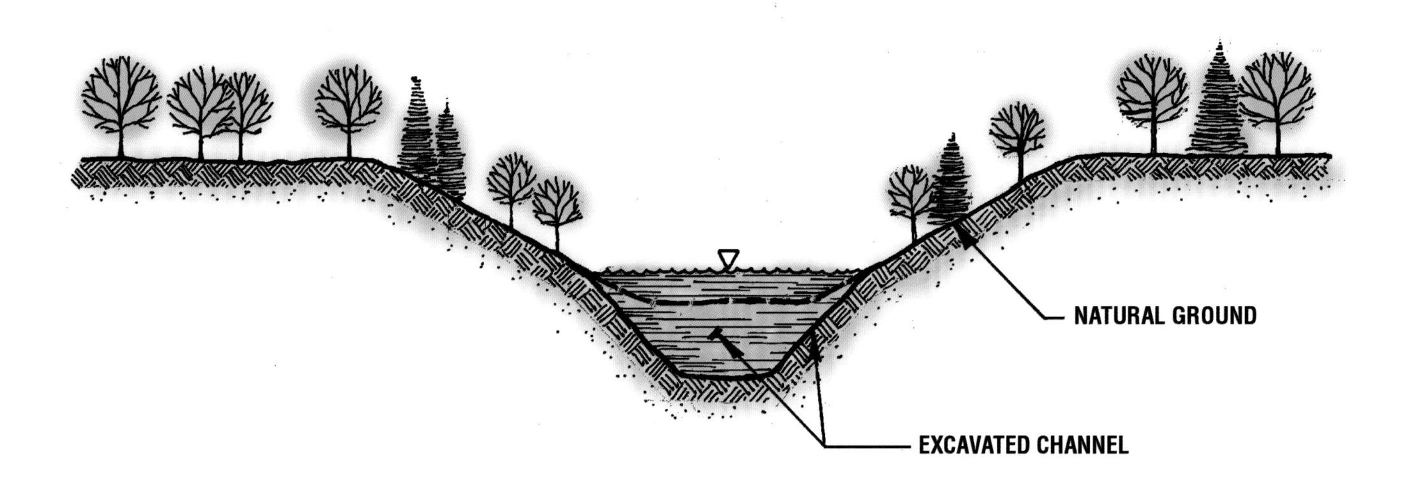

LOCAL

Advantages:

- Increased channel flow capacity
- Increased allowable flow depth
- Reduced flow velocities
- Increased channel water storage
- Decreased water temperature
- Reduced flood damages

Disadvantages:

- Removal of natural streambed cobbles and armor
- Increased erosion
- Possible reduction of bank stability
- Possible degradation of tributaries
- Removal of vegetation
- Disturbance of spawning areas
- Possible lowering of groundwater

UPSTREAM

Advantages:

- Lower flood elevations

Disadvantages:

- Possible streambed degradation

DOWNSTREAM

Disadvantages:

- Increase in sediment loads and decrease in water quality during excavation
- Could increase or decrease peak flow rates

CHANNEL AGGRADATION

FIGURE 58

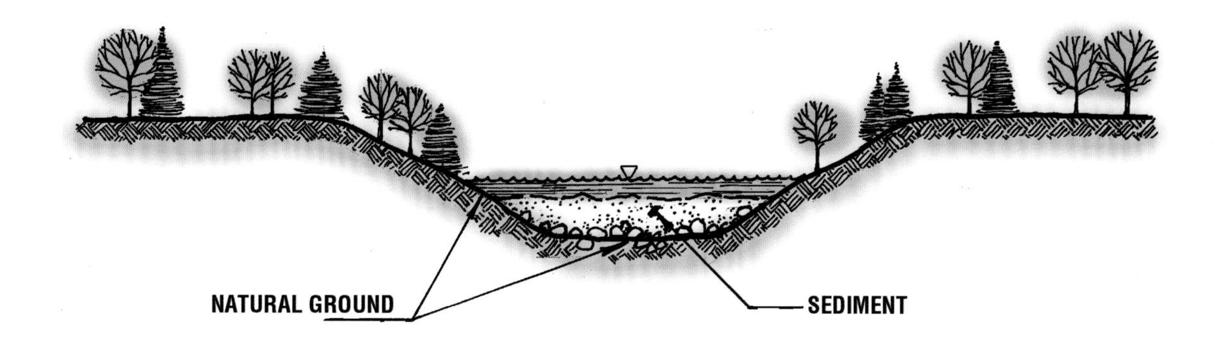

LOCAL

Disadvantages:

- Sediment fills channel bottom
- Filled pools, leading to uniform bed profile
- Destruction of fish habitat and spawning areas
- Reduced flow capacity
- Higher water elevations
- Destroys low flow channels, leading to evenly distributed, shallower low flows, higher water temperature, and degraded habitat

COMMENTS

Channel aggradation is the accumulation of sediments due to excessive sediment loads or insufficient sediment transport. It often occurs during land development when there is insufficient erosion control.

UPSTREAM

Disadvantages:

- Higher water elevations
- Decreased channel slope
- Decreased flow velocity
- Deters fish migration

DOWNSTREAM

Disadvantages:

- Increased initial channel slope and erosion there, increasing sediment load

SAND AND/OR GRAVEL EXCAVATION

FIGURE 59

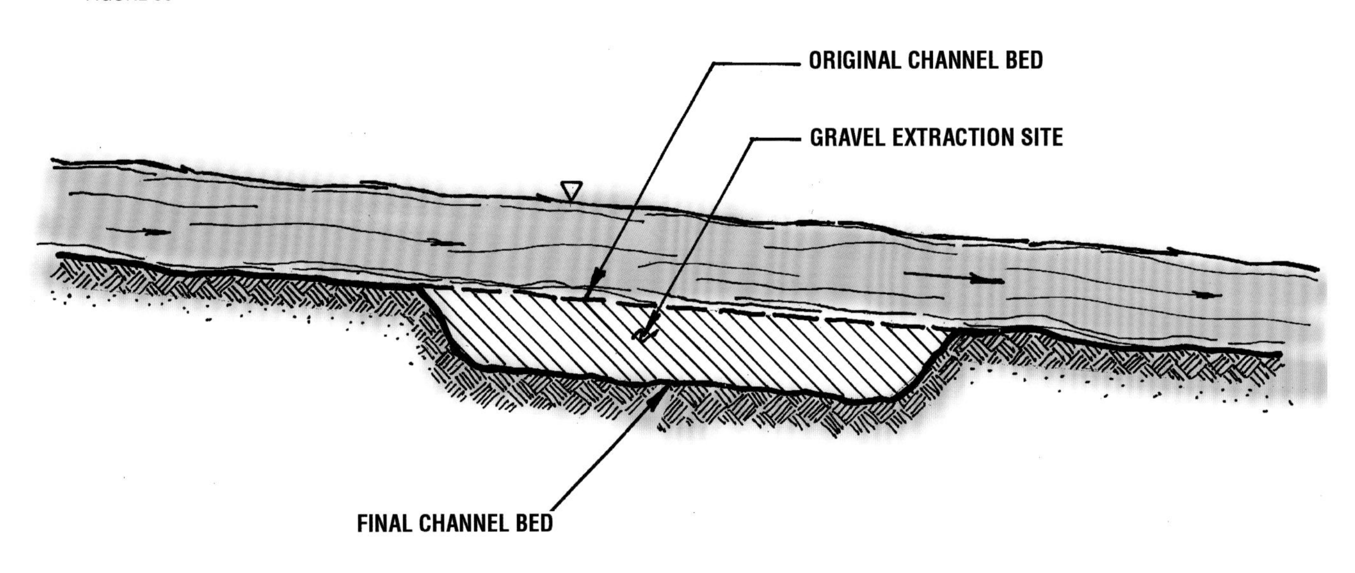

LOCAL IMPACTS

Advantages:

- Source of construction materials
- Possible reduction of flooding during storms of frequent recurrence intervals
- May provide ponds for recreation

Disadvantages:

- Reduced flow velocity
- Increased sediment deposition
- Decomposition of organic sediments may lower dissolved oxygen levels, leading to fish kills
- Steeper and possibly unstable banks
- Reduces substrate variation

UPSTREAM

Disadvantages:

- Potential degradation of channel and tributaries

DOWNSTREAM

Disadvantages:

- Temporary increase in sediment load
- Excavated area may capture sediments, reducing downstream load, causing degradation
- Alters water temperature, dissolved oxygen

COMMENTS

Sand and gravel, deposits of which are formed through the sorting of sediments by flowing water, are resources needed for many types of human activities.

ROAD CULVERT CROSSINGS

FIGURE 60

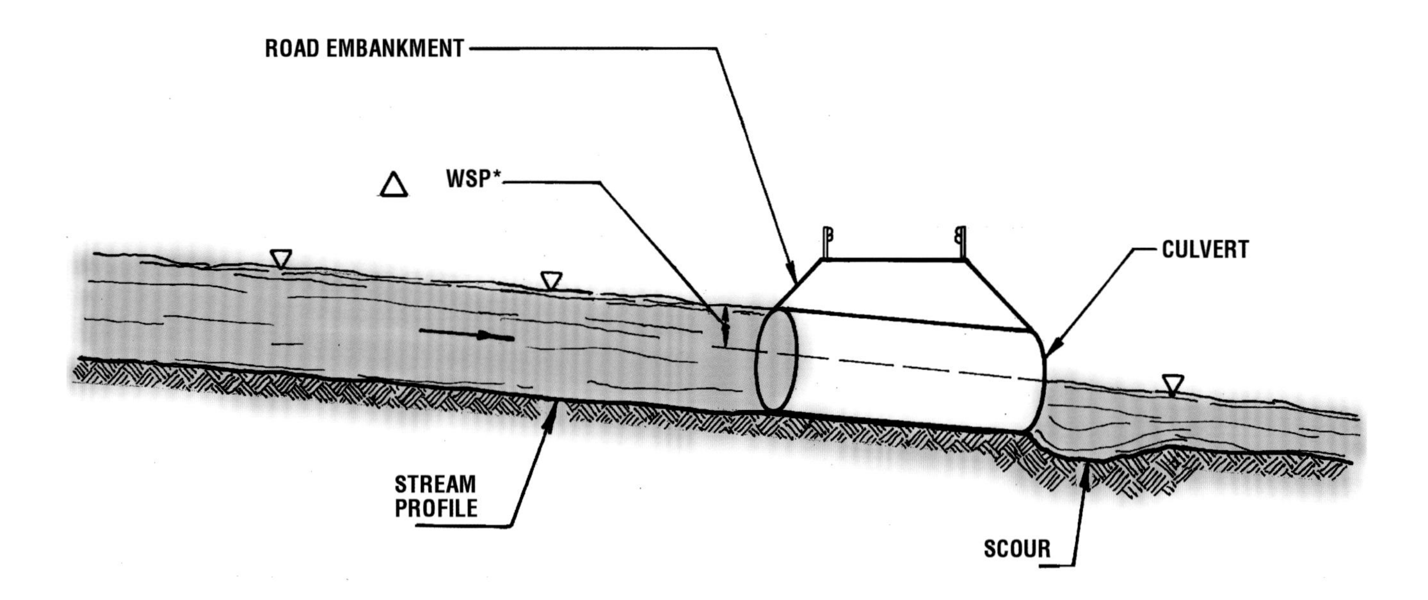

LOCAL

Advantages:

- Access across river
- Lower cost than bridges

Disadvantages:

- Fill constricts channel, reducing habitat area
- Narrows river, floodplain
- Increased flow velocities
- Increased road runoff and debris flow into channel
- Possible obstruction by debris
- Obstruction of fish passage
- Obstruction to boating and fishing

COMMENTS

Eliminates stream habitat.

* Change in water surface profile due to culvert.

UPSTREAM

Disadvantages:

- Raises floodwater levels
- Possible reduction of flow velocities
- Possible increase in sediment deposition
- Potential barrier to fish migration
- Culvert bottom prevents natural bed degradation
- Embankments may act as dam in major floods

DOWNSTREAM

Advantages:

- Ponding upstream of culvert may reduce peak flood flows downstream

Disadvantages:

- Scour due to concentrated flow
- Destabilization of banks

FIGURE 61

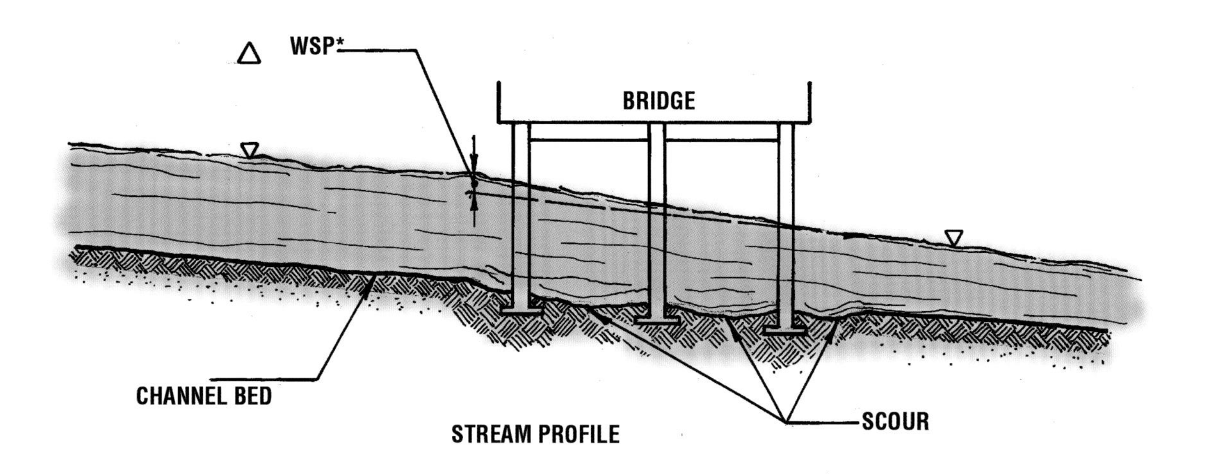

LOCAL

Advantages:

- Access across river
- Less habitat damage than from culverts

Disadvantages:

- Generally higher cost than culverts
- Usually narrows channel, increasing flow velocities and channel scour
- Obstruction to boating, fishing, wildlife
- Possible accumulation point for debris

UPSTREAM

Advantages:

- Usually no barrier to fish migration
- Generally, less damage than from culverts, due to greater flow capacity

Disadvantages:

- Potentially higher floodwater levels
- Embankments can act as dams during floods

DOWNSTREAM

Disadvantages:

- Increased scour

* Change in water surface profile due to bridge.

DAMS

FIGURE 62

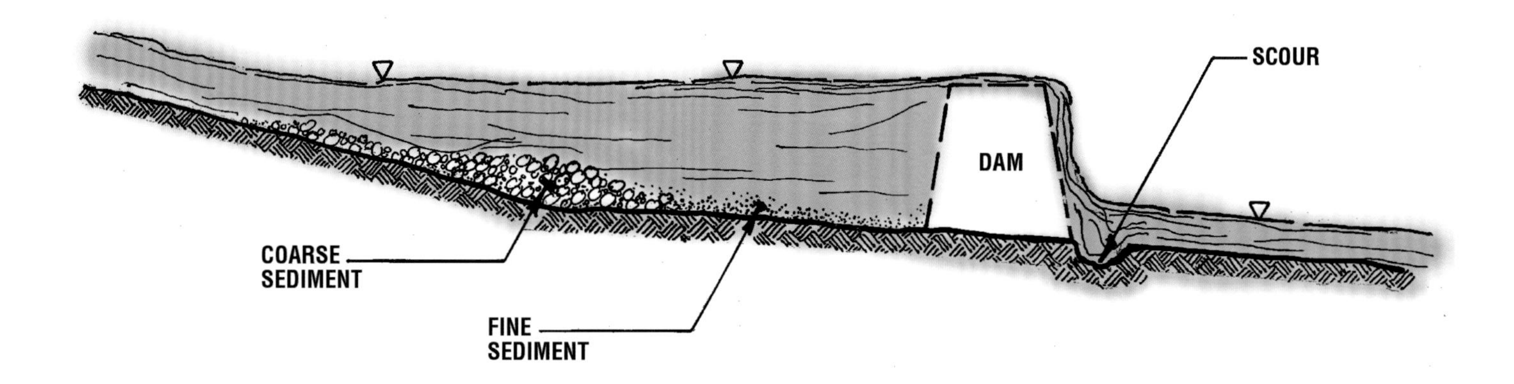

LOCAL

Advantages:

- Creation of lake or pond habitat
- Recreational site
- Potential power-generation site
- Lowers water temperature
- May store runoff
- Stores water for human uses

Disadvantages:

- Decreased turbulence and increased organic sediments lead to less dissolved oxygen
- Increased flood stage
- Floods natural river banks
- Possible flooding of tributaries
- Higher groundwater levels
- Eliminates shallow habitats
- Blocks fish passage, segments river

UPSTREAM

Advantages:

- May provide areas for boating, fishing, swimming

Disadvantages:

- Bed aggradation
- Higher water levels
- Barrier to fish migration
- Fluctuating water level causes bank sloughing

DOWNSTREAM

Advantages:

- May decrease peak flows, increase low flows

Disadvantages:

- Decreased sediment load
- Increased scour
- Dam failures would cause downstream flood with catastrophic flow and scour

COMMENTS

Large dams may provide recreation, flood control and water supply storage. Large impoundments may be used to regulate downstream flow rates.

FLOOD CONTROL DIKES

FIGURE 63

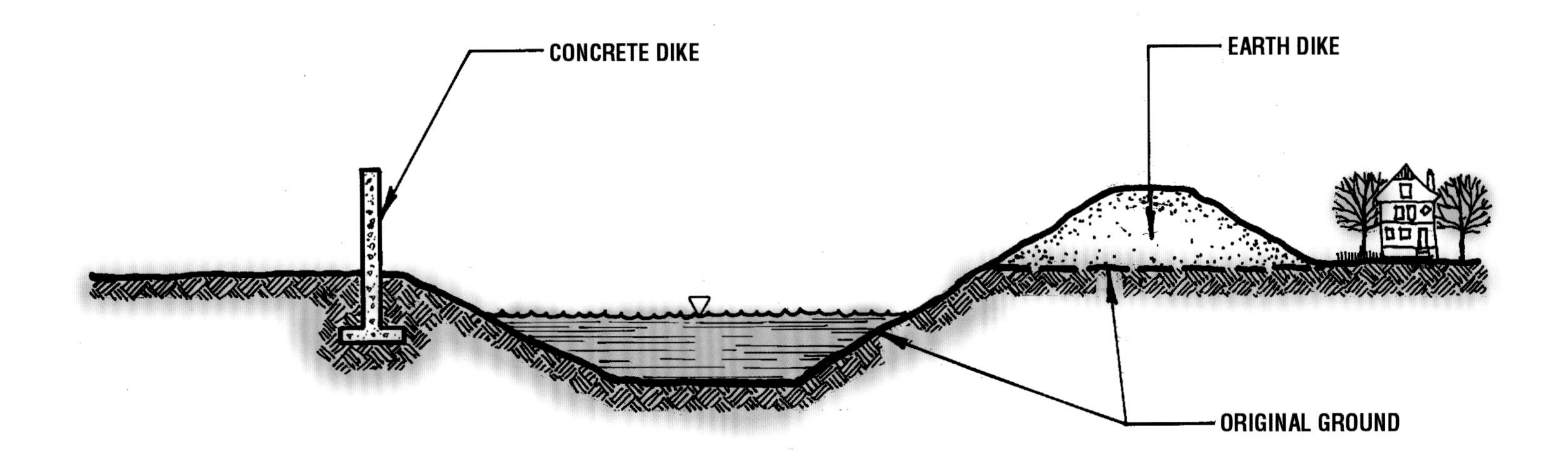

LOCAL

Advantages:

- Prevention of flooding of overbank areas
- Allowance for higher floodwater depths in channel

Disadvantages:

- Increased flood flow velocities
- Increased channel bed scour in low sediment streams
- Increased channel aggradation in high sediment streams
- Increased sediment transport during floods
- Blocking of lateral flow to channel
- Reduced floodplain storage
- Isolation of riparian system
- Prevention of natural meander migration

UPSTREAM

Disadvantages:

- Potential for scour at transition
- Possible degradation
- Possible increase in flood heights

DOWNSTREAM

Disadvantages:

- Potential for sediment deposits
- Scours at transition
- Higher peak flow rates due to loss of upstream flood storage

COMMENTS

Many of the disadvantages of using dikes are reduced when the dikes are set back from the channel, allowing some overbank flow on the floodplain.

RIGID CHANNEL LININGS

FIGURE 64

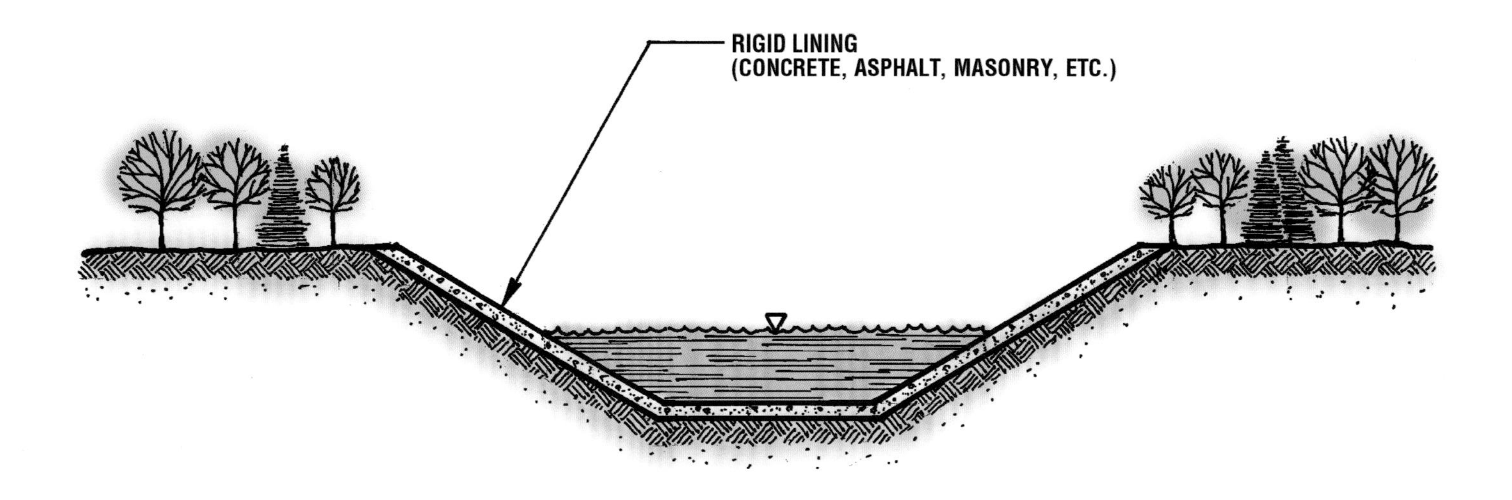

LOCAL

Advantages:

- Prevents erosion
- Low maintenance cost at outset
- Increased flow capacity
- Allows for steeper banks and bed slopes

Disadvantages:

- Increased sediment transport capacity
- Destruction of riparian habitat
- Higher water temperature
- Poor aesthetic appeal
- Lack of habitat diversity
- Decreased channel storage
- Decreased stream-groundwater interaction
- Destruction of fish spawning sites
- Prevention of natural channel adjustments
- Reduces runoff renovation, quality

UPSTREAM

Advantages:

- May lower floodwater elevations

Disadvantages:

- May have high velocities and scour at transitions

DOWNSTREAM

Disadvantages:

- Scour at transition
- Diminished water quality
- Causes more concentrated peak flow rates
- Higher water temperature

COMMENTS

Can create a sterile river with no life.

RIPRAP CHANNEL LININGS

FIGURE 65

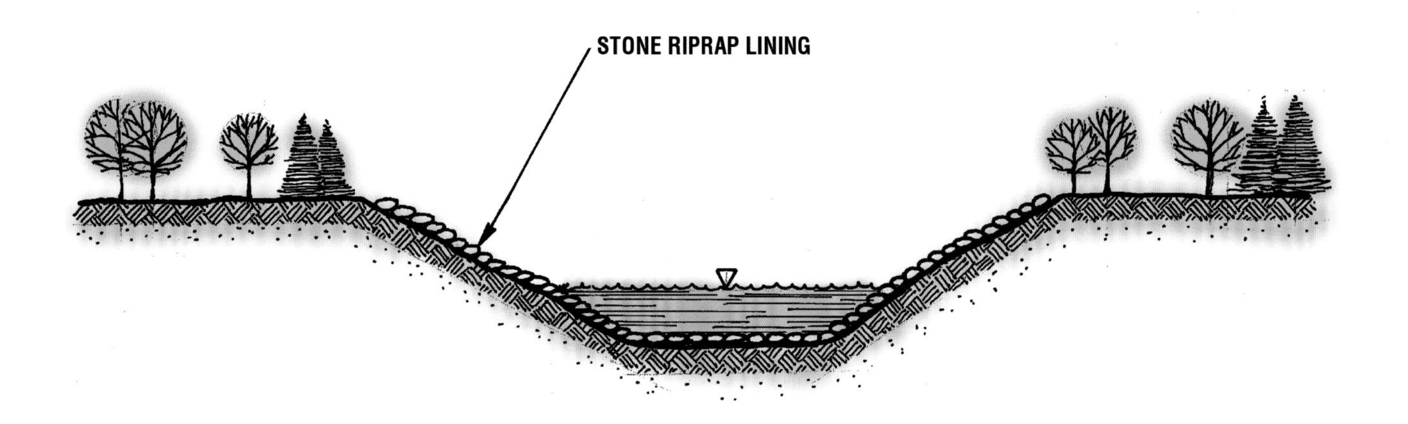

LOCAL

Advantages:

- Minimal erosion
- Increased flow conveyance
- Allows for steeper banks and bed slopes

Disadvantages:

- Prevention of natural channel adjustments
- Pollution from leaching minerals
- Destruction of natural vegetation, wildlife habitat and fish spawning sites
- Poor aesthetic value

UPSTREAM

Disadvantages:

- Scour potential at transition
- Impediment to fish migration

DOWNSTREAM

Disadvantages:

- Increased flow velocities
- Potential for scour
- Higher water temperature

COMMENTS

In areas with low velocities, topsoil can be placed over the riprap and planted with riparian species to mitigate vegetation losses.

STORM DRAIN DISCHARGE

FIGURE 66

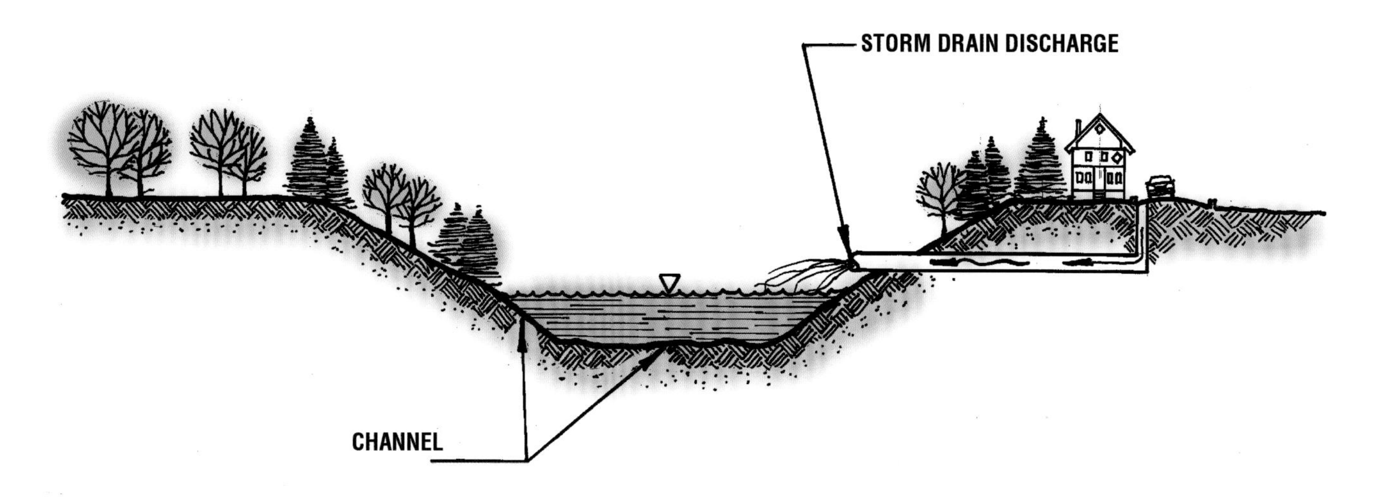

LOCAL

Advantages:

- Discharge point for runoff from developed areas

Disadvantages:

- Possible increase in flow rates in channel
- Increased sediment load
- Potential scour at outfall
- Discharge of pollution

UPSTREAM

Disadvantages:

- Increased flood flow depths if discharge is large

DOWNSTREAM

Disadvantages:

- Increased flood flow rates
- Increased flood flow depths
- Deposition of sediment
- Decreased water quality

COMMENTS

Use of local management techniques to reduce storm drain peak flows and to control runoff quality can mitigate some negative effects.

DETENTION BASINS

FIGURE 67

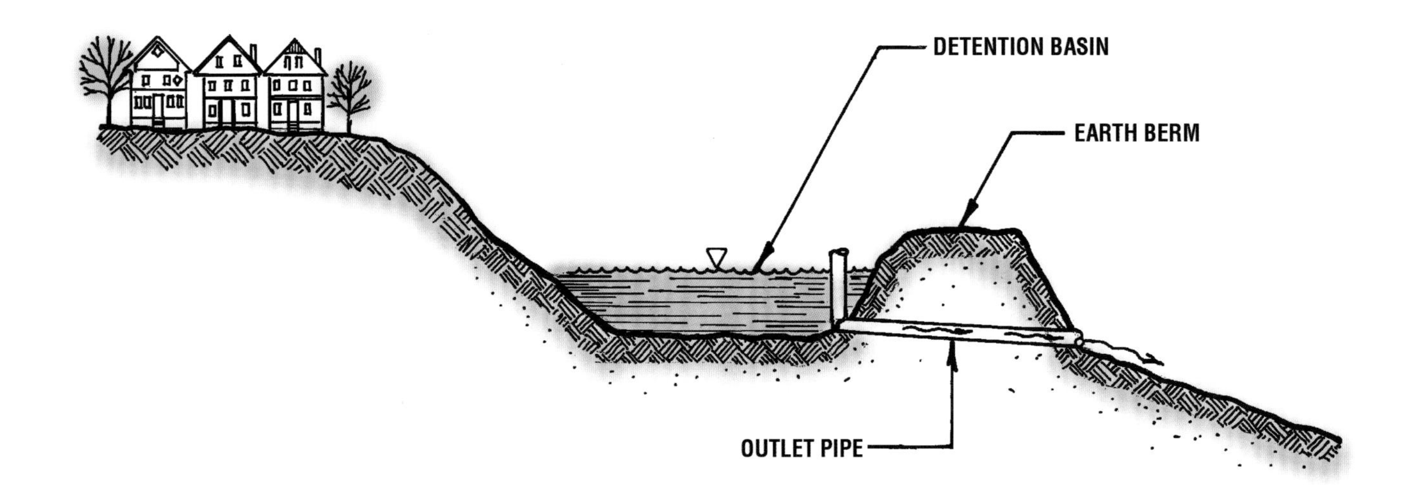

LOCAL

Advantages:

- Temporary storage of excess runoff
- Trap for sediments and urban runoff contaminants, improving water quality
- Possible creation of wetlands

Disadvantages:

- Altered ecology if constructed in natural wetlands
- Limited effectiveness if located in floodplains
- Maintenance required

UPSTREAM

Disadvantages:

- Deterred fish movement

DOWNSTREAM

Advantages:

- Reduction and delay of peak flows
- Possible improvement of water quality

Disadvantages:

- Potential for flooding if dam fails

Comments:

Detention basins and their dams require careful siting, design and construction. The use of offstream detention basins minimizes habitat disturbance. Try to avoid siting them in natural wetlands. Detention basin discharge rates and timing should be coordinated with those of the receiving watercourse.

DIVERSIONS DURING HIGH FLOWS

FIGURE 68

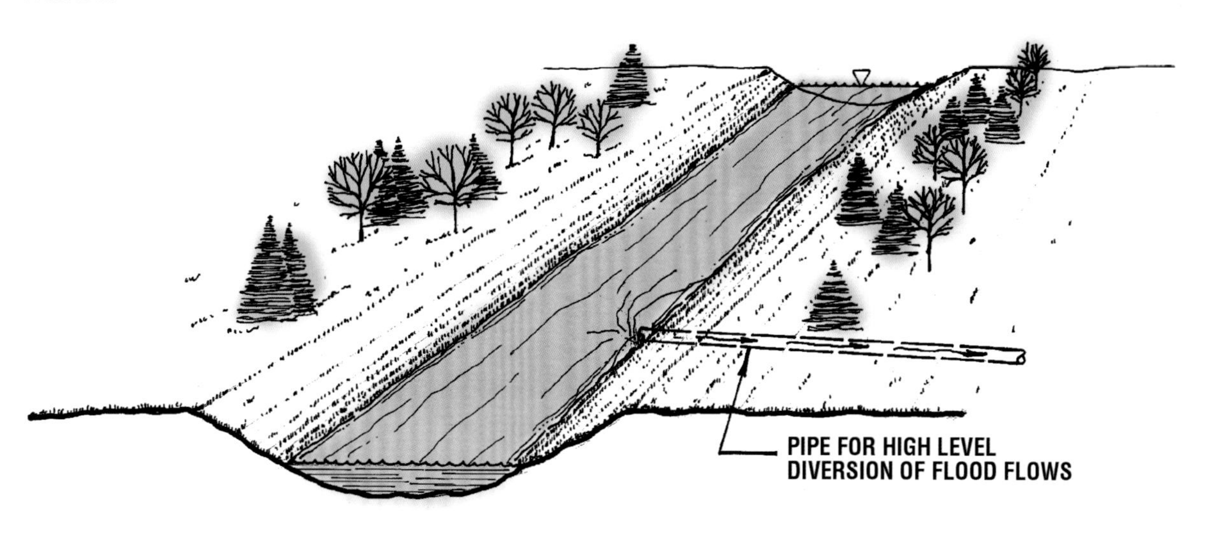

LOCAL

Advantages:

- Reduced floodwater levels
- Reduced sediment transport

Disadvantages:

- Possible change of floodplain enrichment

COMMENTS

Water diversions during periods of high flows help fill reservoirs. Natural high flows are often helpful for removing excess sediment from channels and for fish migration and whitewater recreation.

UPSTREAM

Advantages:

- Reduced floodwater levels

Disadvantages:

- Increased velocity near major diversions

DOWNSTREAM

Advantages:

- Reduced floodwater levels

Disadvantages:

- Less dilution of pollutants, reducing water quality
- Possible degradation of aquatic ecosystem
- Reduced whitewater recreation
- Alteration of sediment transport

DIVERSIONS DURING LOW FLOWS

FIGURE 69

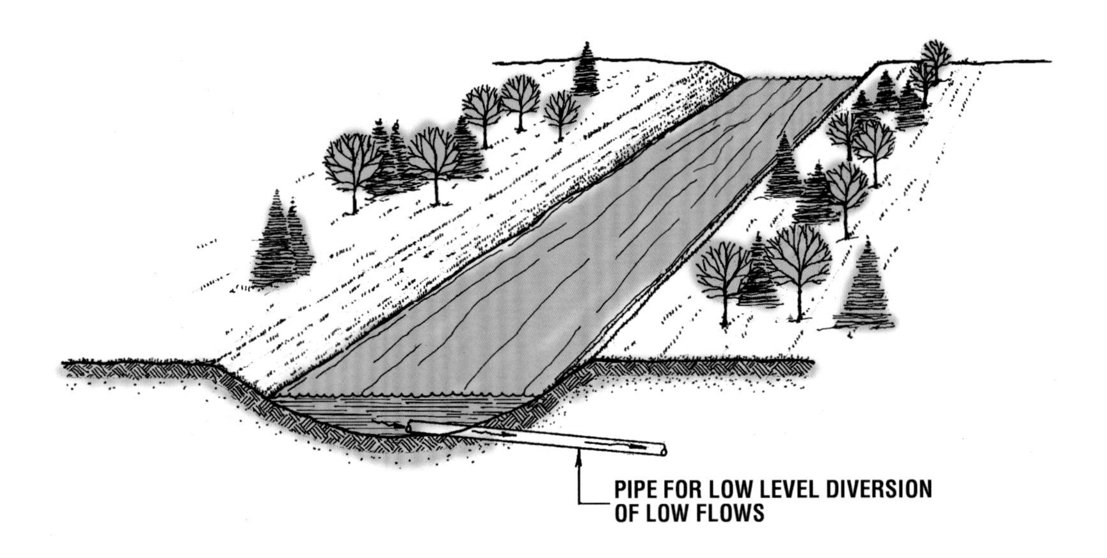

LOCAL

Disadvantages:

- Reduced dilution of wastewater lowers water quality
- Reduced sediment transport leads to aggradation
- Diminished suitability for boating
- Reduced fish habitat area
- Reduced groundwater recharge
- Increased saltwater intrusion upstream in coastal areas
- Potential drying up of wetlands

COMMENTS

Negative impacts are reduced if the water is immediately returned to the river, as at hydroelectric facilities. Long-term diversions during low-flow periods can have a significant impact on the aquatic ecosystems.

UPSTREAM

Disadvantages:

- Fish movement impeded

DOWNSTREAM

Disadvantages:

- Reduced wastewater dilution lowers water quality
- Reduced sediment transport leads to aggradation
- Reduced boating use
- Reduced fish and animal habitat area
- Reduced groundwater recharge
- Increased saltwater intrusion upstream in coastal areas
- Potential drying up of wetlands

FLOODPLAIN ENCROACHMENTS

FIGURE 70

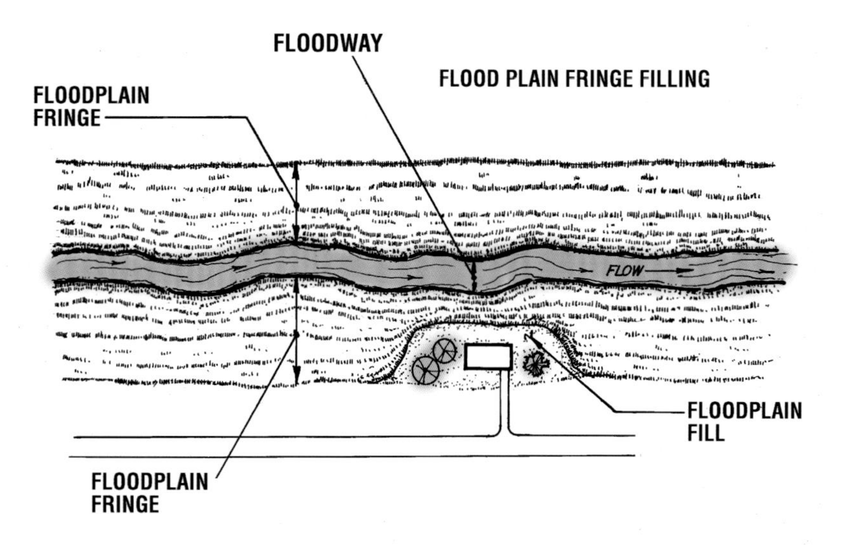

LOCAL

Advantages:

- Reclamation of land for human uses
- Reduced flood damages on filled areas

Disadvantages:

- Increased flood velocities in rest of floodplain
- Increased scour
- Potentially irregular currents and flow patterns
- Destruction of floodplain habitat
- Decreased aquifer recharge
- Possible obstruction of tributary flow to main channel
- Blockage of natural channel meander migration
- Encourages others to place fill
- Reduced floodwater storage

COMMENTS

Floodplain zoning usually allows filling fringe areas. Raising the upstream floodwater profile by the 1 foot allowed by National Flood Insurance Program regulations may cause flood damage to existing buildings.

UPSTREAM

Disadvantages:

- Increased floodwater elevations and flood damages
- Altered flow patterns
- Encourages sediment deposition

DOWNSTREAM

Advantages:

- Possible decrease in peak flood flows

Disadvantages:

- Change in flow patterns
- Higher flow velocities
- Increased sediment load

Public access to man-made canoe channel by-pass

Managing Rivers

> *Learning to read* the Earth and saving it is fascinating stuff.
>
> David Brower, *Let the Mountains Talk, Let the Rivers Run*

FIGURE 71

RIVER MANAGEMENT PROGRAMS

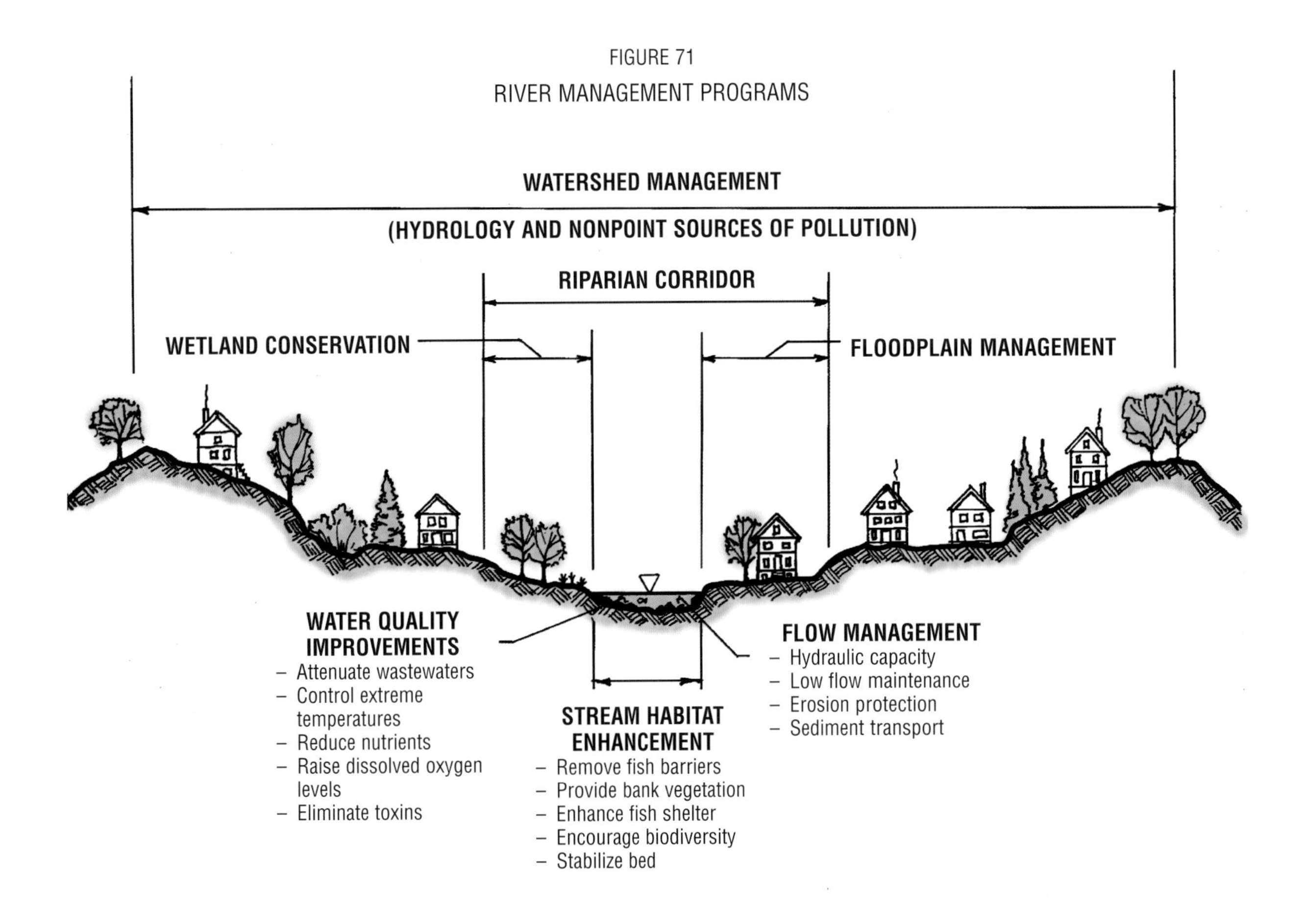

In Connecticut, one is encouraged to contact the River Management and Watershed Management Programs of the Department of Environmental Protection as a starting point for information about Connecticut's rivers.

Managing Rivers

Many factors require that river management efforts extend far beyond the banks that contain the flowing water. Some management issues result from upstream land use, runoff, and sources of pollution. Others arise because of floodplain encroachments, inadequate riparian buffers, or loss of wetlands. The evolving methods of river management emphasize a holistic approach addressing the watershed and stream corridor in addition to the actual channel. There are five broad areas of concern: hydrology, hydraulics, geomorphology, water quality, and the resulting river ecosystem, plus the interrelated social/cultural/recreational uses of the river. (See Table 81.)

Traditional approaches to river management, while often successful, are often limited in scope, prohibitively expensive and environmentally unsound. The concept of managing the watershed and corridor as well as the river channel itself provides an alternate approach that allows each river function to be managed at the appropriate geographic level. (See Figure 82.)

For example, a river's flow rates are dependent on the hydrologic conditions of the entire watershed, not just the narrower channel area. Similarly, a river's water quality depends not only on direct discharges of pollutants, but also on nonpoint sources of pollution throughout the watershed. Consequently, river management must address the entire drainage basin.

The concept of watershed management is consistent with the three-tiered approach that water-supply agencies take in providing potable water: minimizing population densities and sources of pollutants in watersheds, protecting reservoirs from contamination, and treating the water before use.

TABLE 81
WAYS TO PROTECT AND RESTORE CHANNELS

Hydrologic
- Limit impervious cover
- Protect wetlands
- Control peak flows
- Encourage infiltration
- Use natural drainage systems
- Avoid floodplain encroachments

Water quality
- Limit point sources of pollution
- Reduce nonpoint sources of pollution
- Recycle pollutants
- Protect riparian buffers
- Create wetlands
- Limit impervious cover
- Control erosion

Channel Morphology
- Stabilize river banks
- Replant river banks
- Use check dams
- Deflect flow
- Provide low-flow channels
- Avoid channel encroachments

Hydraulic
- Prevent channel filling
- Preserve floodplains
- Minimize channelization
- Use artificial floodways
- Create flood storage capacity
- Provide adequate bridges and culverts
- Regulate floodplain land use

Ecologic
- Preserve riparian vegetation
- Stabilize substrate
- Control sediment
- Provide shade trees
- Provide food sources
- Create pools and riffles
- Preserve detritus sources
- Establish stream corridor greenways
- Remove fish barriers

TABLE 82
GEOGRAPHIC LEVELS OF RIVER MANAGEMENT

ISSUE	WATERSHED	CORRIDOR	CHANNEL
Hydrology	X	X	
Water quality	X	X	X
Normal flow hydraulics			X
Flood flow hydraulics		X	X
Aquatic habitat			X
Riparian ecosystem		X	X

WATERSHED MANAGEMENT

Watershed management has evolved in response to the need for a broad approach that considers rivers to be important natural resources with many, often competing, uses. It is essential to recognize that, besides conveying storm runoff, streams serve many other ecological, economic and social functions, and that the planning and design of management systems must consider water-supply needs, recreational uses, wildlife, aesthetics, and the cost and maintenance of the management measures that are implemented.

Each watershed has unique natural conditions and land uses, and each requires an individual management plan to optimize its sustainable long-term use. The watershed planning process should include local residents as well as local, state and federal agencies. A typical planning process should include:

- Identification of study area.
- Identification and notification of interested individuals, organizations, and public agencies.
- Establishment of an advisory or coordinating board.
- Collection and evaluation of existing natural and cultural features.
- Collection of new data as needed.
- Identification of watershed and stream problems.
- Identification of highest priority issues.
- Evaluation of alternative solutions to problems.
- Researching of funding sources and needed regulatory permits.
- Development of final strategy.
- Preparation of a draft management plan.
- Public review, and possible modification, of draft.
- Adoption of management plan.
- Implementation.

Stormwater Management

If a small watershed is developed in a conventional manner, peak runoff rates will inevitably increase until they exceed the flow capacity of existing channels and drainage structures. The traditional response has been to install larger pipes and channelize the rivers. Recent emphasis on use of detention basins, although sometimes successful, has in other cases compounded flooding through improper release rates and timing, as well as lack of maintenance. Consequently, communities have had to constantly upgrade their drainage systems or to regulate land use and releases of storm runoff to minimize unacceptable increases in peak runoff rates and subsequent channel erosion.

The planning and design of drainage systems must recognize that there are three separate — and sometimes opposing — objectives. The first and most obvious is to provide reasonable drainage facilities for upland communities. This means that drainage plans must protect streets from overland flooding, direct runoff away from buildings, and prevent nuisance flooding at the neighborhood level.

The second objective is to minimize adverse changes in downstream runoff that result from upstream development. A drainage program that conveys water rapidly away might be considered a success in the upland city, but it may increase flood flows, and damages, farther down the watershed.

The third goal is to minimize the environmental and social impact of drainage systems. All too often, they have increased peak flows sufficiently to severely alter the riparian ecosystem, disrupt the stream's stability, and degrade its recreational value.

To create a drainage system that addresses these disparate needs, a comprehensive approach must be used that systematically evaluates the entire watershed. This process requires field inspections and inventories of existing watercourses and drainage facilities, analysis of runoff rates and maximum allowable capacities, identification of existing and potential problem areas, and consideration of future development needs and other social objectives.

Because topography, soil conditions and development patterns may vary in different parts of a watershed, each must be evaluated individually to determine the optimum management plan to be used on the local level. These plans should coordinate the use of natural and man-made drainage systems with the aim of reducing peak runoff rates, delaying runoff that otherwise would be accelerated by channelization, and minimizing the overall increase in runoff volume that urbanization normally causes.

Since the hydrologic impact of runoff from any subwatershed is directly related to its position in the watershed, successful management may require different but coordinated local plans to ensure that peak runoffs from those areas do not converge simultaneously in devastating flood flows downstream.

In existing problem areas, public acquisition of property may be

TABLE 83
MEASURES FOR REDUCING AND DELAYING URBAN STORM RUNOFF

AREA	REDUCING RUNOFF
Large flat roof	Cistern storage
	Rooftop gardens
	Pool storage or fountain storage
	Sod roof cover
Parking lots	Porous pavement
	1. Gravel parking lots
	2. Porous or punctured asphalt
	Vaults and cisterns beneath parking lots
	Vegetated ponding areas around parking lots
	Gravel trenches
Residential	Cisterns for individual homes or groups of homes
	Gravel driveways
	Contoured landscape
	Groundwater recharge
	1. Perforated pipe
	2. Gravel beds
	3. Trench
	4. Porous pipe
	5. Dry wells
	Vegetated depressions
General	Gravel driveways
	Porous sidewalks
	Mulched planters

AREA	DELAYING RUNOFF
Large flat roof	Ponding on roof by constricted downspouts
	Increasing roof roughness
	1. Rippled roof
	2. Graveled roof
Parking lots	Grassy strips on parking lots
	Grassed waterways draining parking lot
	Ponding and detention measures for impervious areas
	1. Rippled pavement
	2. Depressions
	3. Basins
Residential	Reservoir or detention basin
	Planting a high delaying (high roughness) grass
	Gravel driveways
	Grass gutters or channels
	Increased length of travel of runoff by means of gutters, diversions, etc.
General	Gravel driveways

Source: After United States Department of Agriculture, Soil Conservation Service, 1972

TABLE 84

GUIDELINES FOR STORMWATER MANAGEMENT

- Coordinate management of quantity and quality of runoff.
- Use depressions, swales, wetlands and other natural drainage areas to hold stormwater. They provide for a slow release to ground waters (when soils permit). Using these natural storage areas is less expensive than building artificial basins. They also provide recreation, open space, and wildlife habitat.
- Reduce the amount of paved surfaces. Cluster development provides open spaces and infiltration areas, while still allowing for new construction.
- Encourage groundwater recharge with clean runoff where soil conditions are suitable.
- Coordinate stormwater management with erosion control, using wet detention basins with pollution renovation.
- During initial planning, assess runoff flows from areas upstream of the site. Detailed engineering calculations should consider future land uses.
- Manage stormwater so the outflow after development does not exceed the pre-development outflow or stream capacity.
- Coordinate the timing of detention-basin outflows to avoid concentration with peak runoff periods of other watercourses or basins.
- Design drainage systems to include water-quality protection.
- Delineate easements needed for maintenance access.
- Compute runoff routing and storage for two- and 10-year storm events, as well as for 100-year events.
- Assign clear responsibilities for cleaning of filters, removal of debris and sediment, and weed cutting. Consider using a performance bond. Restrictive deed covenants should be used to assure maintenance responsibilities are binding.

Modified from: "Water Quality Guidelines for Development Plan Reviews," Southeast Michigan Council of Governments, 1980.

necessary, or buildings may require flood-proofing or other structural improvements. The management plan should establish priorities and cost estimates for capital improvements such as these, as well as for construction of any new drainage improvements that are needed.

Where development could cause future problems, the emphasis should be on prevention through a combination of land-use planning — including possible limits on some uses — and non-structural flood-control measures.

A comprehensive stormwater management plan that offers the best combination of runoff-control methods and water-quality protection for each part of the drainage basin begins at the local level — that of the neighborhood, and even individual buildings. (See Table 83.) In all cases, efforts should be made to minimize creation of more impervious surfaces and to encourage the infiltration of clean runoff into pervious soils.

On the larger scale, management efforts include measures to reduce runoff volumes, reduce peak runoff, modify runoff timing, and improve flow capacity. Non-structural methods of reducing flood hazards can be especially useful in areas of shallow flooding with low-velocity flow. Many of these techniques are also effective in maintaining the quality of runoff. (See Table 84.)

Some of the most effective methods are those used to keep separate the runoff from the various subwatersheds and to keep them out of phase with each other to minimize the combined peak flow. This prevents the runoff from a river's various tributaries from concentrating at the same time and location. This is accomplished by use of the following principles as illustrated in Figure 72.

1. Controlling the runoff from the lower part of the watershed so that its peak flow does not exceed that of the entire watershed and so its peak flow occurs before the peak runoff from the central part of the watershed reaches the lower part. Generally, this runoff should be passed downstream without delay.

2. The runoff from the central part of the watershed must be delayed until after the peak runoff of the lower watershed has passed, just enough to reduce the peak of the central watershed hydrograph.

3. Runoff from the upper part of the watershed must be delayed long enough to keep its discharge from overlapping with that of the central watershed. A slow, gradual release rate is desired.

TABLE 85

STORM DRAINAGE DESIGN PROCESS

THE INTEGRATION OF SITE DESIGN, DRAINAGE DESIGN, AND POLLUTION CONTROL

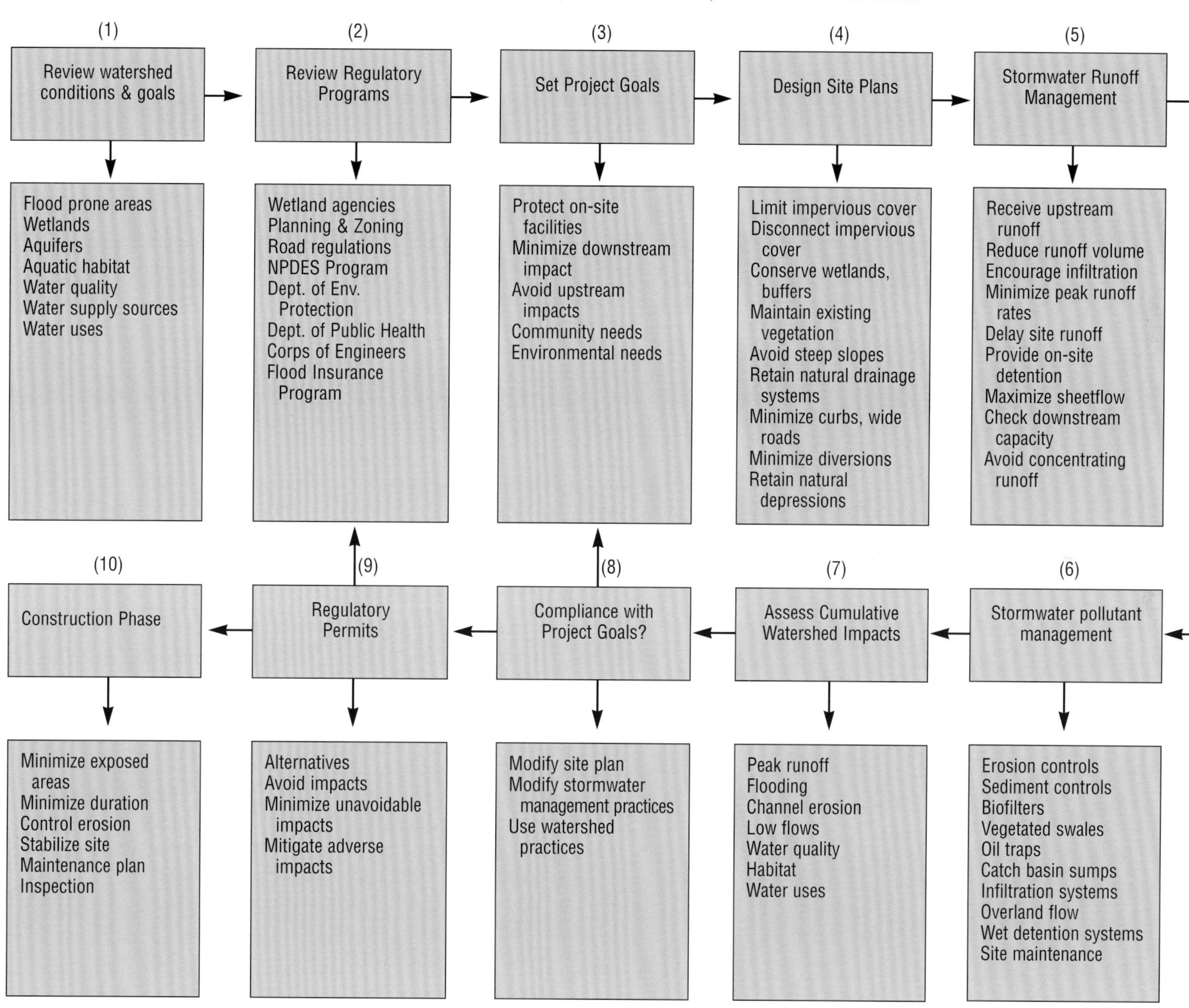

FIGURE 72
WATERSHED TIMING

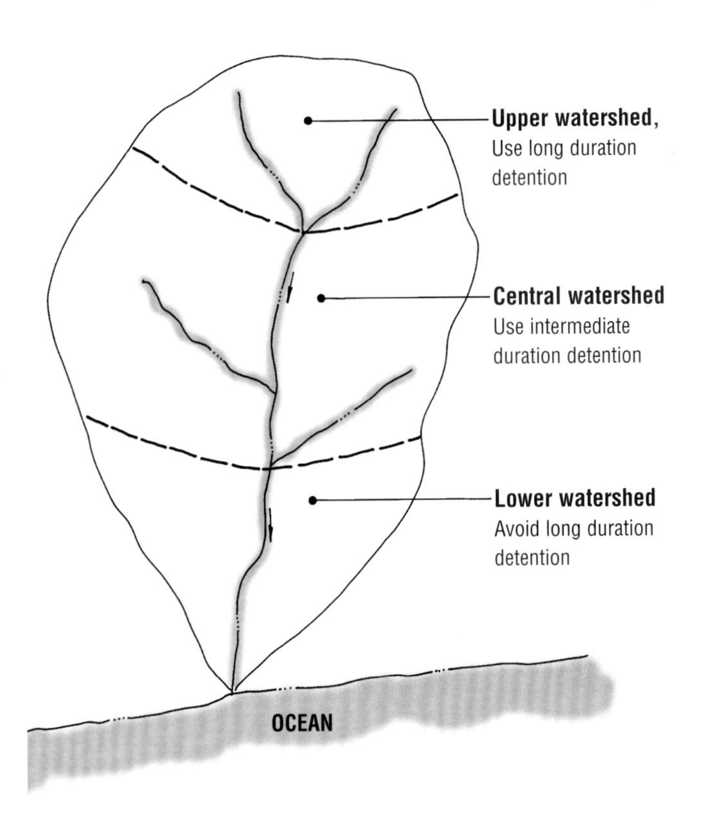

Impervious Cover

Research has found that increasingly large areas of impervious cover in a watershed lead to higher peak flows, larger runoff volumes, channel erosion, poor water quality, and habitat degradation. Negative effects are readily apparent when more than 10 percent of a watershed has impervious surface, while impervious cover of more than 30 percent is associated with severe deterioration.

The watershed management concept includes consideration of upland land use as a river protection tool. Among the methods that may be used to minimize impervious cover in sensitive watersheds are:

- Low-density land uses.
- Use of multistory buildings to reduce surface coverage.
- Cluster development with more intervening open space.
- Minimizing suburban sprawl.
- Permanent open space.
- Preservation of wetlands and farmland.
- Multifloor parking facilities.
- Pervious parking lots.
- Shared parking lots.
- Shared or combined driveways.
- Reduced road lengths.
- Narrower roads.
- Avoiding sidewalk construction in rural areas.
- Use of pervious channel linings.

It also is beneficial to avoid connecting impervious areas directly into drainage facilities, in order to allow runoff to infiltrate into the soil. Specific methods of minimizing the impact of impervious cover include:

- Avoiding use of enclosed drainage systems.
- Directing roof runoff to lawns or woodlands, rather than storm sewers.
- Dispersing road runoff by reducing use of curbs.
- Directing sheetflow and parking lot runoff onto pervious areas.
- Using vegetated buffers around impervious areas.
- Using stormwater infiltration systems to recharge groundwater.
- Preservation of open channels and wetlands.
- Preserving riparian buffers.

The application of stormwater management Best Management Practices, such as infiltration and detention systems, helps to reduce the impact of high ratios of impervious cover on hydrology and hydraulics. Their impact on water quality and biological systems is not fully known.

Stormwater Detention Systems

Recognizing that development can increase the peak rate of surface runoff, some communities have required the use of stormwater detention systems to temporarily store excess runoff, then gradually release it at a reduced rate.

There are many types of systems, including dam impoundments, detention basins, artificial wetlands, underground cisterns, and shallow depressions. They may contain a permanent pool of water or may drain dry between storms, and may serve individual properties or entire watersheds.

For detention measures to work, the following factors should be taken into consideration:

- The discharge from an individual device should be coordinated with that of the overall watershed to minimize simultaneous peak flows.
- Each detention device has a finite storage capacity. Overflow in one could wash out downstream systems, creating increasing flood damage in a domino effect.
- Detention is most effective in the upstream part of a watershed and least effective — or even detrimental — in the downstream end.
- Most devices need a specific program of operations and maintenance if they are to avoid deterioration.
- Channels downstream of detention systems receive flows with reduced sediment levels, encouraging erosion.
- Detention basins should not be placed in natural wetlands that are susceptible to environmental damage.
- Basins should be designed to improve water quality, as well as reduce peak flow, by encouraging pollutant settlement and retaining a permanent pool of water with aquatic vegetation.

TABLE 86

FLOODING TOLERANCE OF SELECTED SPECIES

Less than five days	Less than one growing season
White birch	Bog laurel
Yellow birch	Labrador tea
Paper birch	Black alder
Sugar maple	Red maple
Quaking aspen	White ash
Eastern hemlock	Pin oak
Red spruce	Tamarack
American beech	Atlantic white cedar
Alder	Blueberry
Shagbark hickory	Northern white cedar
Bitternut hickory	Sweetgum
Wild apple	Smooth sumac
Hazel alder	Lilac
Cottonwood	Black gum
Sycamore	River birch
American holly	Hawthorn

Less than thirty days	Less than one year
Walnut	Bald cypress
Basswood	Black willow
Ironwood	Narrow leaf willow
Box elder	Trumpet vine
Apple	Willow
Poison ivy	Buttonbush
Wild grape	Green ash
Virginia creeper	Red-osier dogwood
Red stem dogwood	Deciduous holly
Willow oak	Flowering dogwood
Honey locust	Silver maple
Swamp white oak	Black walnut
Poison sumac	Black oak
Sassafras	Red oak
Black cherry	
White oak	

There is little information on the flooding tolerance of emergent, non-woody vegetation such as herbs, grasses, sedges, rushes, and bog species. Other factors that may be pertinent include plant age, wave action, water quality, and seasonal effects.

(Modified from "Creating Freshwater Wetlands" by Donald A. Hammer, 1992)

Use of Inland Wetlands as Detention Basins

The storage capacity of some inland wetlands is often used to help detain stormwater runoff. This includes the use of natural wetlands and of those that have been modified to increase their water capacity or restrict their discharge rate. Man-made detention basins often overlap wetlands, since the logical location for both is in low-lying areas.

Although natural wetlands often store surface runoff, wetland plants and animals can be harmed by excess water depths or prolonged inundation. In addition, activities intended to increase wetland detention capacities may degrade the wetland. Therefore, such use of wetlands should be avoided where possible.

Generally, using wetlands as detention basins falls into three levels of impact:

Unimproved wetlands

Unimproved wetlands can be used in their natural state to detain runoff if they have an outlet that limits discharges. This usage does not require filling, excavation, or clearing of vegetation.

Modified natural wetlands

The usual modification is construction of a culvert, weir, dam, or similar structure at the downstream end to increase storage volume and retain water. However, increased water elevations can harm wildlife and vegetation, and structures may create barriers to wildlife.

Excavated wetland basins

Excavation is used to increase storage volume where natural low-lying areas do not provide the necessary capacity. This technique is destructive, with extensive removal of vegetation and disturbance of soil, and should not be used in healthy wetland areas.

Discharges of excessive urban stormwater into wetlands can raise water elevations, increasing the frequency of flooding and extending the duration of inundation.

Although vegetation can withstand some variation in water levels, long-term submergence can harm or kill some species, altering the overall habitat. Increased inundation may kill some kinds of animals and force others to relocate. (See Table 86.)

The use of natural wetlands for stormwater detention is appropriate only when the potential impacts on the wetlands' fauna, flora and natural functions are minor. The construction of modified outlets at wetlands must be carefully evaluated to avoid excessive upstream flooding. In general, wetland detention areas that drain within one day should not be harmed.

Instream Flow Management

Whether caused by drought or human consumption and diversions, extended periods of modified flow conditions can subject stream ecosystems to greater stress. Few riverine topics generate more heated discussions than the regulation of minimum flow. Major conflicts can arise between the need for drinking or irrigation water and maintaining flow adequate for aquatic ecosystems and recreation.

The traditional method of establishing instream flow rates for regulated watercourses downstream of dams or diversions is either to stipulate a fixed flow rate or to simulate variable seasonal flows. (Advice is provided by the Connecticut Department of Environmental Protection and the U.S. Fish and Wildlife Service's Aquatic Base Flow Policy.) The following should be considered:

- The needs of migrating fish should be determined, and seasonal flow rates should be of sufficient depth to allow migrating fish to pass over riffles, rapids and low dams and through culverts. Each dam or culvert has to be individually evaluated.
- During its life cycle, each species of fish needs water of a specific velocity, depth and quality.
- Desired flow may vary with water quality and sewage discharges. The ability of a river to assimilate waste changes with temperature, turbulence, oxygen levels, and waste load, so the flow rates should vary accordingly.
- Flow rates and depth needed for boating, fishing and swimming vary from place to place with the river's cross section, depending on channel width, slope and roughness. Consideration should be given for heavy weekend recreation uses and other special activities.
- Flow velocities and rates must be sufficient to meet sediment transport needs and to scour pools. Occasional high flows are desirable.
- Flow rates during the growing season should be adequate to discourage excessive growth of vegetation on temporary sediment bars in the channel.
- Hydroelectric generation schedules and needs should be taken into account, as should downstream diversion needs for water supply, agriculture and industrial uses.
- In coastal areas, river flow should be sufficient to prevent excessive saltwater intrusion into river mouths and estuaries that could affect freshwater species or water-supply intakes.
- Instream flow study methods assess desired flow rates as a function

TABLE 87

RIPARIAN BUFFER ZONE FUNCTIONS

- Maintain water temperatures
- Support and protect fish habitat
- Protect hydraulic capacity of stream and floodplain
- Protect water quality by filtering debris and pollutants from runoff, biological removal of nitrogen, and shading stream from sunlight
- Maintain aesthetic qualities
- Maintain bank stability
- Protect historic and archeological resources
- Prevent floodplain encroachment
- Maintain and protect aquatic vegetation
- Reduce flood-flow velocities
- Prevent overgrazing of banks by livestock
- Recharge aquifers

Simulating a natural channel would be more appropriate than this concrete-lined channel in an urban, riverfront park.

of time. They consider aquatic habitat, fisheries, recreation, water supply, and waste assimilation needs that vary on a seasonal basis.

Water Quality

Watershed management is an important part of maintaining the quality of runoff and the water in streams. The basic elements of water-quality management include reducing sources of pollution, providing adequate treatment of wastes, minimizing pollutant transport, and regulating unavoidable pollutants to ensure assimilation and removal by the river. As discussed in previous chapters, watershed land uses and runoff have a significant impact on a stream's water quality and therefore are important considerations in river management.

Specific watershed measures include controlling soil erosion, treating urban and highway runoff to remove pollutants, minimizing agricultural runoff, preserving wetlands, treating point discharges of pollution, and maintaining adequate streamflow to assimilate treated effluent and unavoidable discharges. Methods of treating point discharges of pollution and controlling nonpoint sources are presented in Chapter Four.

RIPARIAN CORRIDOR MANAGEMENT

The riparian corridor includes lands adjacent to the river that have a direct and immediate impact on it. The corridor, which should be managed in combination with the channel itself, is the primary area of interaction between the aquatic and terrestrial habitats. (See Table 87.)

The width of a stream corridor management area varies with local conditions. It should include the floodplains and wetlands connected to the stream, plus land areas that contribute organic detritus and overland runoff directly to the stream. Tributaries have their own corridors, forming a branching network within the watershed.

The watershed management measures discussed previously are also applicable within stream corridors. Stormwater-quality management and erosion control are especially important because of the close proximity to the channel.

Besides the stream itself, the focus of corridor management programs is generally on preservation of vegetated buffers, wetlands and floodplains, and on creation of water-related greenbelts and urban riverfronts that have high recreational and other social value.

Stream Buffer Functions

The preservation of undisturbed vegetated buffer zones along rivers is effective in sheltering rivers and wildlife from short-term construction activities and long-term land uses such as agriculture or urban development.

The buffer areas help to filter runoff and trap sediment, provide wildlife cover, shade the water, and limit vehicular access. Buffer-zone widths should be based on local conditions, including soil types, slope, vegetation, runoff rates, and runoff quality.

Riparian zones improve the quality of runoff through physical, vegetative and soil processes. (See Table 88.)

The physical factors include filtration and settlement that reduce the amount of suspended sediments reaching the river in runoff, plus the adsorption of dissolved materials from runoff.

Bank vegetation along lakes and streams shades the water, helping limit temperature changes. The root zone and forest litter help to stabilize the soil, retarding bank erosion.

Through its root system, vegetation takes up nutrients, helping to store them during the critical growing season, and slowing the rate of nutrient transfer to the river and its transport downstream. Microbes in the root zone convert some nitrogen, a major component of treated septic wastes, into a gaseous form, removing it from the aquatic system, where it contributes to excessive algae growth. Such denitrification is common in peat, muck and carbon-rich soils.

The effectiveness of riparian buffers varies with several factors. Narrow buffer zones have less opportunity to renovate upland runoff, and provide less habitat area. Buffers that slope steeply toward the stream channel will have higher runoff velocities, reducing the zone's efficiency in trapping sediments. Low-permeability soils of silt and clay reduce the buffer's groundwater recharge capacity, and thin vegetative canopies provide less shelter and shade.

Buffer Widths

Most of the scientific literature supports basing buffer widths on site-specific, functional criteria such as hydrogeologic and cultural features, including topography, soil types, present or desired water quality, existing riparian-zone development, upland land use, and stream-management objectives.

The most important component of a stream buffer program is

Table 88
RIPARIAN BUFFER IMPACT ON
NONPOINT SOURCES OF POLLUTION

Pollutant Sources
- Prevents pollutants from being stored, applied or discharged near river's edge.
- Reduces soil erosion.

Pollutant Mobilization
- Vegetative canopy reduces raindrop force.
- Forest leaf litter helps to shelter pollutant particles.
- Microtopography reduces flow rates.
- Surface roughness reduces flow velocities.
- Vegetation disperses runoff.

Pollutant Transport
- Microtopography reduces flow rates.
- Surface roughness reduces flow velocities.
- Pervious soils reduce runoff through infiltration.
- Irregular topography increases travel distance and time.

Pollutant Attenuation
- Dilution with clean water.
- Adsorption on soil, vegetation and litter surfaces.
- Low velocities encourage settlement.
- Soil and vegetation filter pollutants.
- Biological uptake of soluble materials.
- Shade reduces water temperature.
- Snags trap sediments.

TABLE 89
RIPARIAN BUFFER LIMITATIONS FOR
RENOVATING NONPOINT POLLUTANTS

- Can be overwhelmed by high runoff rates
- Channel formation limits attenuation
- Soil has limited adsorption capacity
- High groundwater limits infiltration
- Nutrients are released by plant decomposition
- Floods can resuspend sediments
- Steep slopes have higher flow velocities
- Limited impact on dissolved solids

TABLE 90

TYPICAL ALLOWABLE RIPARIAN BUFFER USES

- Low-density residential lots
- Tree farms
- Selective timbering
- Recreational facilities
- Existing agricultural activities

POTENTIALLY HARMFUL RIPARIAN BUFFER USES

- Sewage-disposal systems
- Salt storage sites
- Storage or use of hazardous wastes
- Industrial facilities
- Pesticides, herbicides
- Unfenced grazing
- Clear-cutting vegetation
- Manure storage
- Mining activities
- High-density residential dwellings
- Large parking lots
- Sanitary landfills

preservation of the first 25 feet of land near the edge of the water. This zone, which should be highly protected, includes the streambank, canopy trees that overhang the stream, and aquatic vegetation along the water's edge. The supplemental, or outer, buffer zone, often as wide as 100 or 200 feet, provides additional protection, while still usable for low-impact human activities. (See Table 90.)

The U.S. Army Corps of Engineers has found that buffer strips of 30 meters or less are adequate for many protective functions. For example, small streams require 10- to 20-meter buffer zones for adequate shade. A 20- to 30-meter-wide buffer has been found to be adequate to filter sediment from surface runoff as long as the flow across the buffer zone does not become channelized. A 20-meter buffer has successfully reduced nitrogen levels in both surface water and groundwater.

A 1993 U.S. Environmental Protection Agency manual, "Guidance Specifying Management Measures for Sources of Nonpoint Pollution in Coastal Waters," recommends "that no habitat-disturbing activities should occur within tidal or nontidal wetlands. In addition, a buffer area should be established that is adequate to protect the identified wetland values. Minimum widths for buffers should be 50 feet for low-order headwater streams with expansion to as much as 200 feet or more for larger streams. In coastal areas, a 100-foot minimum buffer of natural vegetation landward from the mean high tide line helps to remove or reduce sediment, nutrients, and toxic substances from entering surface waters."

Standards developed by the U.S. Forest Service for design, operation, and maintenance of forested riparian buffers to protect water quality in the eastern United States recommend a minimum of 75 feet from the streambank.

A national survey of 36 stream buffer programs by the Metropolitan Washington Council of Governments found urban stream buffers from 20 to 200 feet in width, with a median value of 100 feet. Buffer-zone widths may be adjusted to include wetlands, floodplains, and steep, erosion-prone slopes.

The following Best Management Practices are suggested for managing buffer zones:

- Avoid clear-cutting of timber.
- Avoid major land clearing for agriculture, construction, development, or other activities.
- Avoid extensive use of heavy machinery, which can cause ero-

sion by breaking down fragile streambanks and by compacting the soil so that it can no longer absorb runoff. Soil compaction also prevents aeration of the soil, which, in turn, causes vegetation to die and eventually leads to erosion.

- Where any land clearing does occur in the buffer, re-establish native grasses, shrubs and trees as soon as possible.
- Except at stream crossings, avoid building roads in the buffer zone.
- Avoid building near a streambank. During floods or heavy rains, disturbed banks can slump many feet in a short time.
- Land use within buffer zones should be regulated to permit activities consistent with protecting the stream environment, while discouraging activities that cause pollution or degrade the stream corridor. Since most riparian areas are private property, it is necessary to allow reasonable uses.

Several factors limit the effectiveness that stream buffers have in renovating stormwater runoff, so it is necessary to apply Best Management Practices in regulating activities in the buffer zone. (See Table 89.)

This sterile concrete channel, built in the 1960s, has a high flow capacity but low environmental value.

This multi-use, man-made flood control channel, on the same stream as shown in above photo, was constructed in the 1980s with a bicycle path.

Greenways

Open-space corridors, or "greenways," along streams are a valuable and effective method of conserving river resources.

Greenways are public and quasi-public rights-of-way that form linear corridors along streams. They are frequently used to connect larger parks and open-space areas. Unlike buffer zones, greenways are usually available for limited public uses such as hiking, biking, fishing, and similar activities. In urban areas, even though there may be fewer natural features, they provide recreation and access to the river.

Rural greenbelts are used heavily by wildlife. They provide nesting areas, shelter and cover, and sources of food. In addition, they provide travel routes between undeveloped areas. (See Table 91.)

The New York Department of Environmental Conservation distinguishes between stream corridors, and objectives for their management, according to the amount of nearby development, as follows:

Urban These are waterfront areas characterized by hard surface paving, storm-drainage systems, and land uses related to commerce and industry. The overriding concern is to take advantage

TABLE 91
GREENWAY USES

• Open space	• Parks
• Playgrounds	• Urban walkways
• Picnic groves	• Bikeways
• Nature trails	• Canoeing
• Wildlife habitat	• Aesthetics

of the opportunities that the corridor can provide to enhance livability in the city. Imaginative use of corridors can provide significant economic and recreational benefits, such as parks, historical sites, and focal points for redevelopment.

Suburban These have a combination of vegetated ground covers and pavement, storm drains and natural drainage, and a wide range of land uses including residential, commercial, and industrial. Here managers should establish a greenway to serve as a filtering and buffering area to protect and enhance water quality. Passive recreation — such as sightseeing, nature photography, and observing wildlife — and active recreation, such as hiking and fishing, are appropriate activities. Planners should preserve historical sites along such a stream corridor.

Natural These are greenways with cultivated land and natural vegetation, with little pavement. Farming and recreation are the prevalent uses. The primary emphasis should be preservation of open space and protection from agricultural runoff. The buffering or filtering capacity of the natural streambank should be maintained to protect water from effects of mining, logging and other industrial processes that may take place upslope.

Urban Rivers

Restoring and maintaining the health of urban waterways is critical in densely populated regions like Connecticut. Not only can a river provide a valuable recreational and economic resource for city residents, it can also help to increase awareness of issues related to water quality, which in turn can aid in the overall effort to safeguard the health of Connecticut's waterways and Long Island Sound.

During the 20th century, American cities turned away from their riverfronts. Waterfront factories, warehouses, railroad tracks, highways, and flood-control dikes contributed to isolating entire neighborhoods from their rivers, which often were little missed because water quality was so poor. In some cases, entire watercourses were buried in underground channels.

Recapturing the river's edges for public parks can be very costly. But restoration of urban rivers and their waterfronts can be an important part of improving the urban landscape and the quality of life in towns and cities, as well as reducing flood damage.

Urban rivers will not offer the same benefits as rural rivers with their natural environment. But with more emphasis on visibility and human use, an accessible and attractive urban riverfront provides many benefits. (See Table 92.)

While preservation may be the goal along rural rivers, a premium should be put on public access and use of urban riverfronts. It is essential to attract people and activities into river parks to create interesting and safe areas.

In suburban areas, streams are apt to be the last natural element in an otherwise artificial environment. Even though surrounding areas are likely to be developed, suburban floodplains still provide valuable wildlife habitat. People enjoy being in such areas, making them prime locations for low-impact recreation.

Archeological and Historic Sites

Many Native American settlements were established along riverbanks and on floodplains. Similarly, early European colonists usually settled near rivers. Many historic sites are there to be unearthed and highlighted as cultural attractions.

Before construction is begun along rivers, state archeologists should be contacted to see if a site has such historic value.

Excavation of Sand and Gravel

Sand and gravel deposits are a non-renewable resource often located along channels and under adjacent floodplains. These materials are used in construction projects or as an aggregate in concrete and asphalt. Mining sites must be carefully planned to minimize riverine impacts. (See Table 93.)

Removal of the material can be done either above the water table or below it. The first type of excavation is less harmful than the latter. Both, however, can alter the shape of the floodplain, causing water flow to be diverted during periods of flooding. There have been cases in which deep excavated ponds have accumulated contaminated sediments that otherwise would have been carried downstream. Decomposition of organic sediments in gravel ponds may reduce oxygen levels and degrade fish habitat.

Accordingly, sand and gravel extraction must be carefully monitored to prevent accidental flow diversions, flooding, and water-quality problems.

TABLE 92

URBAN RIVERFRONT BENEFITS

- Provides open space
- Offers attractive views
- Provides recreational opportunities
- Boosts retail and commercial activity
- Promotes environmental awareness
- Fosters neighborhood pride

TYPICAL USES

• Marinas	• Walking trails
• Boat rides	• Flower gardens
• Historic sites	• Frequent/visible access points
• Cultural events	• Zoos or aquariums
• Sporting events	• Fishing platforms
• Holiday celebrations	• Viewpoints
• Band shells and amphitheaters	• Natural exhibits
	• Water taxis

TABLE 93

SAND AND GRAVEL SITE MANAGEMENT PRACTICES

- Provide buffer zones between rivers and excavated pond
- Preserve riverbank vegetation
- Minimize area of disturbed soils
- Do not stockpile sand and gravel in floodways
- Replant disturbed areas
- Avoid steep subsurface slopes
- Do not excavate near bridges, dams or dikes
- Avoid dredging during fish spawning and migration periods

Gravel excavation in a major urban river. Note the collapsed bridge span in the background.

In some cases, natural or man-made dikes may be needed to separate an excavation site from the river.

When extraction ends, closure plans should be prepared to ensure restoration and reuse of excavation sites. For example, some sites have been converted into public parks.

National Flood Insurance Program

The National Flood Insurance Act of 1968 was passed to provide insurance for flood-prone areas and to reduce flood damages. In order for its residents to qualify for federally backed flood insurance, a community must adopt special land-use and building regulations for its floodplains. The program is thus an important part of river management.

The Federal Insurance Administration, a division of the Federal Emergency Management Agency (FEMA), has conducted studies for every town in Connecticut. They document peak floodflow rates, floodwater elevations and floodplain boundaries.

FEMA has set minimum standards for areas regulated under the Flood Insurance Act. Participating communities may elect to adopt more stringent rules.

Key floodplain management elements include:

- No hydraulic constrictions can be installed within the inner floodplain corridor (floodway) of a 100-year flood.
- All floodplain activities are limited so as to prevent the base floodflow elevation of a 100-year flood from rising more than one foot.
- All new residential buildings must have their lowest floor above the elevation of the 100-year base flood.
- All buildings must be able to resist flotation and the pressure of floodwaters on their outer walls.
- Utility lines, pipes and other conduits must be located so as to minimize damage.
- The lowest floor of all new nonresidential buildings must be elevated or the building floodproofed.

Although the Flood Insurance Act offers a comprehensive plan for reducing flood damage, it does have limitations.

The program uses historic floodflow data and data current when the community chooses to participate, and does not provide for any future increase in peak flow rates or water elevations caused by development upstream. Therefore, future floodwater elevations may be higher than those cited for regulatory purposes, especially in small watersheds sensitive to land development.

Further, standard computations for floodplains (see Figure 73) focus primarily on flow capacity and do not take into account the loss of water-storage capacity that occurs when the fringe areas are filled or developed. This means that floodplain development permissible under these regulations actually can increase peak flow rates more than anticipated and raise water levels above the allowable one-foot increase for a 100-year flood. Because the FEMA program allows for a one foot rise in the elevation of the base flood, it is wise to always design and build to at least one foot above the base flood.

Consequently, new buildings constructed in areas the regulations consider to be above the base flood level may actually be subject to future flooding.

The scale of maps used for planning generally ranges from one inch equals 400 feet to one inch equals 1,000 feet. Such scales are not always sufficient to accurately delineate limits of the floodway and floodplain.

Some hazard areas are not studied in detail, but are mapped as being in an "A" zone. While some limits are placed upon land use within these areas, there are no detailed engineering computations on peak flow rates and flood levels. Developers in these areas are expected to prepare their own flood studies and submit them for approval.

The program was developed primarily for insurance purposes. Consequently, it concentrates on mapping major watercourses in communities where development has already occurred. Yet watercourses too small to be mapped can still produce significant flood damage to newly developed areas within the stream corridor or to existing development downstream.

Criteria used in determining whether fringe areas can be filled or otherwise altered are based on potential hazards. To that extent, the program implies that such fringe areas can be filled. However, some consequences of such activity are difficult to measure in engineering terms alone.

For instance, many fringe areas contain valuable habitats or rare or endangered species. Often, they are important groundwater-recharge regions that help infiltrate surface water into the underlying sand and

FIGURE 73

FLOODPLAIN ENCROACHMENTS AND FLOOD ELEVATIONS

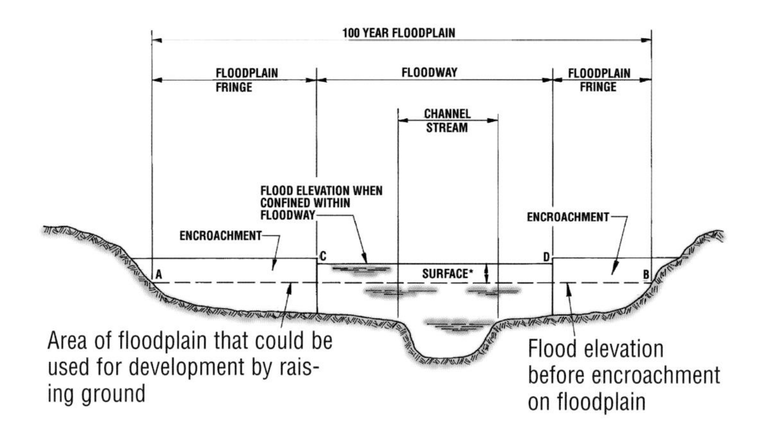

Line AB is the flood elevation before encroachment
Line CD is the flood elevation after encroachment
*Allowable increase in flood elevation is not to exceed 1 foot (Flood Insurance Act requirement) or lesser amount if specified by state

gravel aquifers. The insurance program, established to minimize flood damage, is not designed to protect other riverine resources. Local officials, property owners, and others evaluating proposals impacting flood flows often have available to them other, more stringent, requirements for protecting riverine environments.

Maps showing hazard areas imply that the floodway boundary is a fixed line. In actuality, the main channel and floodways of many rivers can be altered during the course of a flood as a result of factors such as scour, debris or ice jams.

The program is implemented by local commissions rather than on regional watershed bases. This can mean inconsistent and conflicting policies within the same watershed.

New Jersey is one of several states that has recognized the limitations of the federal Flood Insurance Program standards and has adopted more stringent requirements. They include:

- Defining floodways to allow an increase in the floodwater profile of only 0.3 foot (versus the federal one-foot rise). Effectively, this requires a floodway wider than the federal standards call for.
- Defining flood fringe area, based on a flow equal to 125 percent of the flow that could be expected from a 100-year flood, thus allowing for the runoff associated with future development.
- Restricting filling to 20 percent of the total fringe area.

Meander Belt Hazard Areas

Points of land between meander bends are among the most active floodplain areas, and are subject to frequent inundation. Several of Connecticut's most flood-prone areas are such meander-enclosed locations: the Glastonbury meadows of the Connecticut River, the Indian Field area of New Milford along the Housatonic River, and sections of Farmington and Simsbury along the Farmington River.

Unfortunately, the flood insurance studies generally do not address possible channel realignment on floodplains. Meander belts are, in fact, hazard areas that should not be developed.

CHANNEL MANAGEMENT

The alteration of natural watercourses should be avoided where possible. When natural channels must be disturbed and when degraded channels are restored, modifications should be made in a manner that causes the least environmental damage while meeting the community's drainage and flood-control needs.

All channel alterations must be planned carefully, so that the new channel has adequate hydraulic capacity, is dynamic over short-term periods yet stable in the long term, and is attractive and environmentally sound.

At the outset, planners should inventory the existing and potential characteristics of the natural river. The research team should include experts in engineering, geology, biology, ecology, and land-use planning.

For any given site, the important factors are:

- Areas subject to flood damage
- Erosion and sediment deposition
- Water quality and aquatic habitat
- Wetlands
- Vegetation
- Wildlife
- Soils and geology
- Aesthetics
- Recreation potential
- Land use

After the inventory, the team must identify areas of special interest. It then must decide what improvements should be carried out.

Each river is unique. Each requires a solution that is unique.

Creation and Restoration of Stream Channels

The following management techniques are applicable to the creation of new channels and the restoration of existing streams, encompassing hydraulic, morphological and ecological concerns.

Hydraulic Analysis

Hydraulic analysis of open channels is performed to determine their flow depth, velocity and capacity.

Analysis of irregular and non-uniform channels is complex. One needs to consider energy losses due to friction and at bends, contractions, and expansions; the effect of sediment on friction; the irregular geometry of bed and floodplain; secondary currents; flow separation at bars or islands; and differences in velocity at different points in the cross-section of the channel and floodplain.

Sometimes a portion of the floodplain may be inundated but —

FIGURE 74

DUAL CONVEYANCE SYSTEM

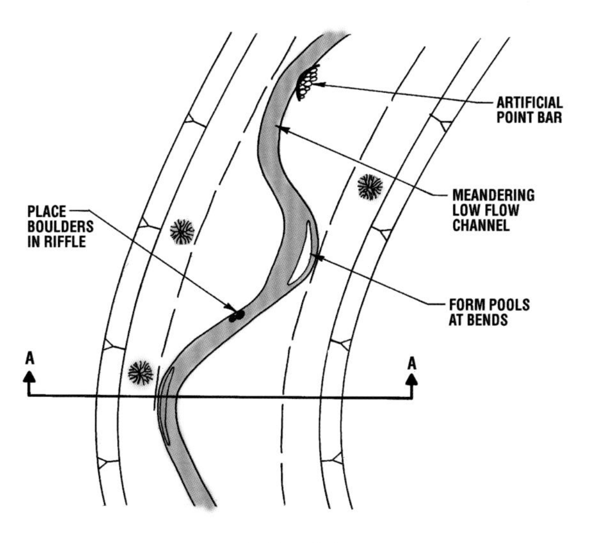

Section A-A

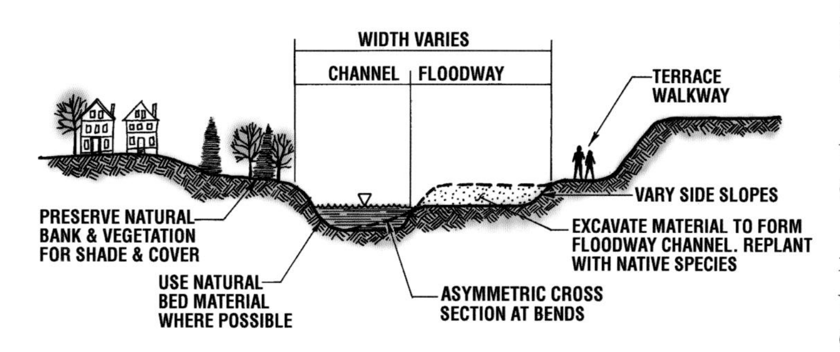

because of its shape, location or roughness — does not actually convey water. Such ineffective flow areas must be identified along with areas of effective flow.

Most hydraulic computations for rivers with complex geometry are done with computers. Many of the programs are limited to linear flow in rivers with rigid boundaries, with no scour or deposition, and with constant friction that does not vary with channel width or depth of flow. New two-dimensional programs are coming into use that will be more accurate. Friction factors should be noted for the entire cross-section. Values should be obtained for at least three segments, representing the channel itself and the floodplain on either side. They should assume summer vegetation conditions.

Hydraulic analysis of new channels should consider downstream water elevations, alignment, width of floodplain, losses of energy, friction, downstream clearance for ice and debris, bridges, embankments, and the effect of utility pipes suspended beneath bridges.

Channel Capacity

Man-made channels usually are designed to convey large flood flows, such as that of a 100-year event. However, they often are unstable because they are excessively wide and shallow, and have little cover, providing poor aquatic habitats.

One way to minimize these problems is to construct a channel with two elements, a base channel for normal flows and a floodway for excess flows.

The base flow capacity of man-made channels should approximate natural equilibrium conditions. Natural channel stability typically occurs when the capacity equals the flow of a 1.5- to 2.5-year flood. Flows in excess of these rates are normally conveyed by the floodplain. If the latter is absent or is inadequate because of development, a floodway may be built to form a dual conveyance system. (See Figure74.)

The ideal channel would be relatively narrow, compared with those conventionally used for flood control, with sufficient flow depth to transport sediment and support fish, generally maintaining a base flow one to two feet deep.

After setting the size of the main channel to carry normal flows, the floodway is proportioned for the larger floodflows. The surface of the floodway should be higher than the main channel; usually dry, it can be planted with ground cover and occasional trees. It could also form a greenbelt corridor that would serve as a buffer between the main channel and urbanized land, while also being used for recreation. The floodway could be on either or both sides of the main channel, with a variable width.

In cases where encroachments must be made on the natural floodplains, it is desirable to maintain their conveyance capacity by adjusting the floodplain cross section. This may not be possible in areas that are already heavily developed, but should be a priority where new development is occurring.

The left bank of this man-made floodway is constructed to carry floodwater while minimizing disturbances to the channel. The right bank is kept in natural condition to provide wildlife cover, shade and aesthetic variation.

Alignment and Slope

The appropriate combination of channel alignment and slope should be based on the principles of fluvial morphology.

A man-made channel should replicate the bed slope and discharge rates of a stable natural channel in order to minimize erosion and the need for artificial linings, and to provide favorable aquatic habitat.

The most desirable patterns are slightly sinuous, avoiding the instability associated with braided channels.

A meandering pattern is attractive in that it is conducive to a wide range of habitats. However, it may use a lot of land and cause scour problems at bridges. A straight pattern requires little land and is easy to construct, but offers less opportunity for having self-scouring pools.

Sometimes, where the slope is severe, it may not be possible to create the desired pattern. In these cases, the bed may be lowered by using drop structures, small dams across the channel that create a steplike profile.

Cross Section Design

The shape and size of channel cross sections must balance hydraulic and environmental needs. The channel's average depth and width are set after determining the channel's natural capacity and slope. The required cross section geometry is based on hydraulics for a given situation. Cross sections that are compact minimize the perimeter, and thus the costs and amount of land needed.

The width-depth ratio should be similar to that of a healthy, natural river. The width and depth should vary to provide adequate flow conveyance while ensuring a diversity of aquatic habitats. The side slopes may vary, providing diverse habitats and an aesthetically pleas-

A dual conveyance channel with a low-flow channel in the center.

ing appearance.

The optimum channel width is the maximum size for the desired flow depth without encouraging low velocities and formation of excess sediment bars. If the channel is too narrow, it will be subject to excessive scour and degradation. If too wide, it will tend to collect debris and have shallow flow unsuitable for fish. It should have a well-defined thalweg.

Sometimes a channel changes in width, due to irregular flow patterns, erosion, deposition, and energy losses. The design must take into account these transition sections. At points where the channel contracts, velocity increases. Therefore, the channel must be deeper at such sites or must be lined to resist erosion. At points of expansion, velocity is slower, encouraging sediment bars. Here it must be made shallower.

Channel Stability

Whether a channel is alluvial or non-alluvial, the ability of bed and bank material to resist movement should be carefully analyzed.

The design of channels with high sediment loads must consider sediment transport requirements. The sediment load is first measured or estimated, and sediment-transport formulas are then used to determine the channel size required to maintain a dynamic stability. Such methods for estimating the size of alluvial channels are useful guides, but must be combined with engineering judgment and experience.

Fortunately, Connecticut watercourses generally have low sediment loads because of the dense vegetation and glacial-till soils. Many channels in the state have essentially rigid banks and semi-rigid beds.

In clear-water channels with little sediment load and in non-alluvial rivers, the type of watercourses that exist in many of Connecticut's upland areas, channels should generally be designed for non-scouring flow conditions.

Channel Profile

The optimum man-made channel should include a series of pools and riffles with a meandering thalweg. This will re-create the irregular geometry of a natural river without having to wait for the new stream to establish its own micro-patterns.

These elements should correspond in size and proportion to those of natural rivers. The spacing between pools should be 10 to 14 times the channel width. Pools should be at the outside of the meander bends. The U.S. Soil Conservation Service recommends that at least 35 percent of a stream's length consist of riffles, with another 35 percent being formed of pools, the latter with no overall slope.

In straight channels, a meandering thalweg may be created, and stones may be used to alter the bed elevation to create riffles where desired.

Controlling Aggradation

Aggradation occurs when the sediment load exceeds the stream's transport capacity, filling the channel with deposits. The process can be controlled by reducing the load or by increasing the transport capacity.

The optimum approach is prevention: instituting good land-use policies that discourage soil erosion. Sediment traps, debris basins and vegetated buffers can also limit the amounts of sediments reaching the stream. These measures can be applied readily in small watersheds but are difficult to implement in large ones.

Sediment transport capacity can be increased in several ways. The preferred method is to make the river narrower, forcing flows to converge and concentrate, creating higher velocities that can carry more sediment. Flow velocity can also be increased through increasing the slope of the bed by straightening the channel, or through reducing hydraulic friction by removing obstructions such as boulders, snags and vegetation. However, these steps may have negative environmental impacts.

Sediment transport may also be increased through the use of dikes, flow deflectors, groins, and similar structures to create turbulence that keeps sediment in suspension.

Controlling Degradation

Degradation occurs when the rate of scour and sediment transport exceeds sediment supply. It may be controlled by reducing transport capacity or increasing erosion resistance.

The transport rate can be lowered by decreasing peak flow rates, by decreasing the channel slope using a series of sills or weirs, by widening the channel, or by increasing its length by establishing a more sinuous pattern. The last two options are practical only when enough land is available. One preferred method of controlling degradation is to use natural materials to increase flow friction and reduce

TABLE 94

CHANNEL RESTORATION REQUIREMENTS

Hydrology	Control/reduce peak floodflow rates
	Ensure ample in-stream base flows
Hydraulics	Adequate channel/floodplain capacity for peak flows
	Avoid high velocities
	Increase channel roughness
	Provide diverse flow types and velocities
Morphology	Stabilize eroding bed and banks
	Provide proper width/depth ratio
	Use appropriate pattern and profile
	Confine in-stream flows to discrete channel
	Balance sediment scour/transport
	Maximize coarse substrate
	Remove debris
Water quality	Treat point sources of pollution
	Control/minimize nonpoint sources of pollution
	Reduce watershed sediment sources
	Provide shade, minimize solar exposure
Aquatic habitat	Revegetate banks
	In-stream habitat diversity
	Requires coarse and fine organic matter sources
	Remove fish barriers

velocities to non-erosive levels. This may include planting the banks with water tolerant species, placing boulders along the bed, and anchoring logs along the banks and bed.

In severe cases, another way of controlling degradation is lining the channel with material that resists erosion. This includes rock riprap, concrete slabs, retaining walls, and gabions of small rock in wire-mesh baskets. However, these methods negatively affect the ecology and restrict access to the stream.

Channel Restoration

Where streams have been disrupted by human activities, generally in developed areas, restoration efforts may be undertaken to revive a channel's flow efficiency, aesthetics and environmental values without extensive new work. Proper restoration helps to maintain the channel and stabilize it without extensive straightening, widening or deepening.

Such projects are often accompanied by floodplain management so as to reduce the need for flood-control channelization. In urban areas, restoration may be aided by programs that control erosion and sediment movement, and that discourage any increase in peak rates of runoff.

For larger restoration projects, concerns also include controlling peak flows, maintaining adequate low flows and improving water quality. (See Table 94.)

The restoration process should also include removal of man-made obstacles, including abandoned culverts, bridges, piers, utility pipes, and fences. (See Figure 75.)

Channel Linings

Artificial channel linings are commonly used to minimize bank and bed erosion. They protect the bank from being washed away and improve water quality by reducing sources of sediment. Smooth linings decrease flow resistance, thereby increasing flow and capacity.

Such linings are usually associated with one of the following:

- Flood-control projects
- Bank repairs after floods
- Bridge or culvert crossings
- Storm-drain outlets
- Land development
- Erosion control

FIGURE 75

TYPICAL STREAM RESTORATION MEASURES

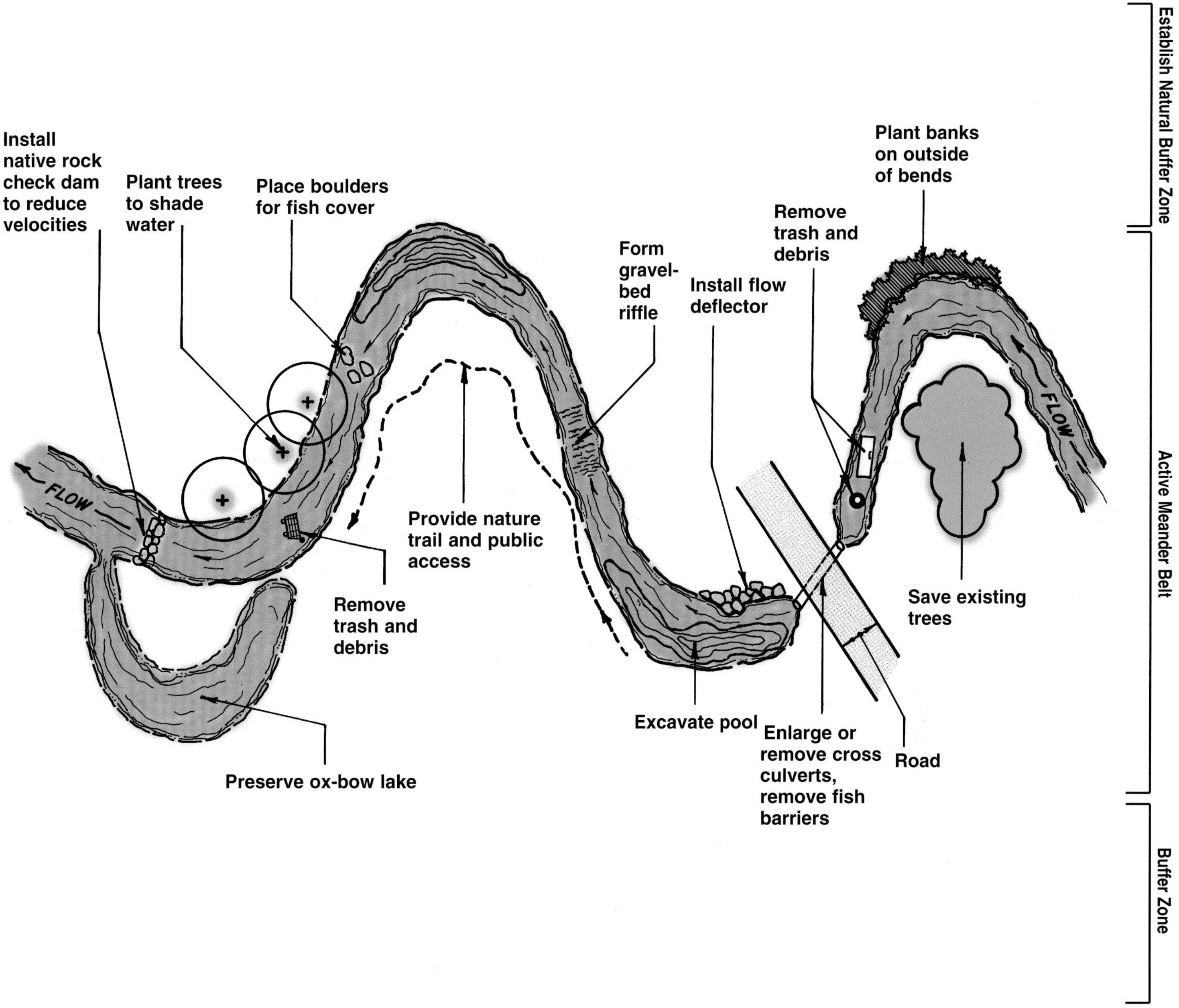

Install native rock check dam to reduce velocities

Plant trees to shade water

Place boulders for fish cover

Form gravel-bed riffle

Install flow deflector

Remove trash and debris

Plant banks on outside of bends

FLOW

FLOW

Provide nature trail and public access

Remove trash and debris

Save existing trees

Preserve ox-bow lake

Excavate pool

Enlarge or remove cross culverts, remove fish barriers

Road

Establish Natural Buffer Zone

Active Meander Belt

Buffer Zone

TABLE 95

POTENTIAL IMPACT OF ARTIFICIAL CHANNEL LININGS

- Loss of wildlife cover and habitat
- Loss of shade trees, increased water temperature
- Unattractive appearance
- Poor fish habitat
- Disruption of normal sediment balance
- Increased downstream velocity and erosion

TABLE 96

METHODS OF CHANNEL LINING

- Vegetative ground covers
- Stone riprap
- Soil bioengineering
- Precast concrete blocks
- Gabion baskets filled with stone
- Soil concrete
- Synthetic fabrics
- Concrete slabs
- Stone masonry

Channel linings can have negative impacts, however, especially where they are used in long reaches. (See Table 95.)

Accordingly, an artificial channel should be used only when no other option is feasible. If one must be built, careful evaluation of floodflows, velocities, bank and bed erodibility, and long-term ecological conditions is necessary in order to select the most appropriate lining. (See Table 96.)

Vegetative Ground Covers

Vegetative ground covers are commonly used where there is low erosion potential, for instance in a channel with low flow velocities, erosion-resistant soils, and mild slopes. Ground cover has less deleterious impact than does stone, concrete or artificial materials. Vegetation allows infiltration to occur and helps retard the flow of floodwaters.

Such covers may be established by seeding or placement of sod. It takes more than one growing season for vegetative covers to become established, and they may need supplemental seeding.

They are not effective below the normal water level, however, and are prone to damage by heavy foot traffic or if they are buried by sediment deposition.

Riprap Linings

Riprap, consisting of broken stone, is a common material for lining channels. The size of individual pieces should be large enough to resist the velocity of the flowing water and its tractive force. Sizes vary from a few inches to several feet in diameter. Occasionally, even larger rock is used in areas subject to severe turbulence or wave action.

Since it cannot be rolled by the current, broken stone from quarries or construction sites is superior to rounded stone. Where rounded rock is used, it must be large enough to resist high-velocity flows. Hard, durable rock, free of fractures, is preferred. Rock that is soft or subject to splitting, such as some limestones and sandstones, is undesirable. Unweathered stone may leach soluble minerals into the water.

Excessively large or deep riprap can be detrimental on small streams, because it allows low flows to pass through the voids, eliminating the riparian habitat and effectively transforming a perennial stream into an intermittent one. Unfortunately, several formulas used to determine the appropriate size of stone necessary to resist movement have been found to be flawed.

The depth of riprap layers should be twice the thickness of the larger stones in order to ensure good coverage. Riprap should be underlain by gravel bedding or geotextile fabric filters to prevent erosion of soil.

There are several methods to reduce the negative impact of riprap linings:

- Place the riprap below the river bed and place native bed sediment over it.
- Make the riprap surface irregular with pools to encourage bed variation.
- Place bank riprap deeper than normal; cover it with topsoil and replant it.
- Preserve vegetation on the banks.
- Preserve large shade trees.
- Minimize riprap on the channel bottom, limiting use to side slopes.

Soil Bioengineering

Soil bioengineering (also called biotechnical methods) has been used in Europe for many years and is becoming more common in the United States. These techniques make use of both ecological and engineering concepts to stabilize slopes, embankments and riverbanks.

Cuttings from live shrubs, saplings, and even mature trees are planted in slopes during the dormant season to reinforce the earth. Such plantings intercept rainfall, encourage infiltration, reduce flow velocities, and minimize erosion. The subsequent root growth binds the soil together.

Live cuttings can be planted directly into the the riverbank, or layers of cuttings can be placed so that they project from fill material. Bundles of dead brush cuttings can be interspersed to retard erosion until the live cuttings grow and fill in the gaps.

The vegetation employed must be able to grow in a wet environment, produce new roots rapidly, and form a dense layer. Desirable sources include willow, cottonwood, sycamore, some red maples, and forsythia. Use of native plants is recommended to help match habitat and climate conditions.

These techniques may be combined with conventional plantings of rooted stock.

The advantages of these methods are that they employ local mate-

A man-made dam creates a pool along this small river.

rials and are compatible with natural conditions. The disadvantages include high flow friction that may raise water elevations locally and upstream, and high labor costs.

ECOLOGICAL CONSIDERATIONS

Aquatic life, which depends for survival both on the quality of stream channels and the quality of the water, should be a major consideration when planning changes to a watercourse. (See Table 97.)

It is important to have proper design for the overall channel and floodplain system to ensure structural and environmental stability. Constructing new channels to simulate natural channels is an effective way to reduce and partially compensate for the ecological damage fostered by river projects.

Reduce ecological losses

Minimize potential damage by avoiding stream-corridor disturbances where possible.

Compensate

Offset any unavoidable damage by creating new habitats.

Restore and enhance

Protect disturbed habitats from further damage and improve them where possible; reverse destructive activities.

A riverine environment is subject to constant change and stress. Although some species are able to adapt to change, extreme alterations undermine the larger ecological system. A watercourse is considered altered when sediment, water quality, water quantity, hydraulic characteristics, or vegetation is seriously disturbed.

High peak flows scour bottoms and banks, removing vegetation, food sources, and shelter. Many species cannot tolerate such rates for long. Conversely, a flow that is abnormally low, such as the reduced base flows often found in urban areas, may have low oxygen levels and may not be able to transport sediment. Deposition can bury organisms and prevent the formation of a stable bed.

Natural variety of conditions provides the desired diversity of habitat. For instance, trout require cold water with shaded gravel beds in riffles. But they use deep, slow moving pools for shelter during hot weather and during high discharge periods. Other species prefer warm water with sandy bottoms.

At moderate velocities, scouring may benefit a habitat, since it removes fine grains from coarser soils and contributes to formation of desirable gravel beds. However, excessive scour can wash out both gravel bottoms and vegetation. High flow also can remove organisms from the bed, reducing food for fish and waterfowl.

It is critical to maintain sheltered areas along a river. These can be in the form of logs, stumps, boulders, overhanging vegetation, snags, or man-made flow deflectors that serve to trap organic material and provide hiding places.

Ideally, riffles should be composed of fine and coarse gravel. They may also be cobbles or bedrock. Water depths should be at least six to 12 inches, with flow velocities of two to three feet per second.

Vegetation along banks is important for shade. For rivers more than 30 feet wide, mature, overarching trees should line the banks along at least half of the river's length.

Instream structures can create varied flow conditions, with microhabitats that meet the various needs of many aquatic species. But such measures are of little benefit on streams with poor water quality or a lack of basic nutrients. Therefore, control of both point and nonpoint sources of pollution is also critical. (See Table 98.)

Fish Passage at Culverts

Drainage culverts that convey streams beneath roads, parking lots and other facilities often deter movement of fish and other aquatic species. Their passage can be inhibited by high flow velocities, lack of resting areas, poor light, and harsh materials.

Culverts can be designed so as to encourage passage both upstream and downstream. To minimize obstruction, the following features should be considered:

- Use open-bottom culverts with natural bed material.
- Use oversized pipes with the bottom buried below the grade of the natural streambed.
- Avoid high-velocity flow, especially where long culverts are used.
- Provide high culverts with air space and light above the normal water level.
- Place baffles along the culvert bottom to reduce velocities and provide resting areas.
- Design the culvert elevation, width and slope to avoid excessively shallow flow.

TABLE 97

MEASURES THAT PROTECT AN AQUATIC HABITAT

Maintain vegetated buffer zones.

Buffer zones help filter runoff and trap sediment. The vegetation shades the water, minimizing temperature rise, and maintains organic input to stream. The Connecticut Department of Environmental Protection Fisheries Division recommends a 100-foot buffer along perennial streams, and a 50-foot buffer along intermittent ones.

Monitor groundwater diversions.

Shallow wells that tap into the groundwater supply should not be dug along streams that have low rates of flow. They may reduce streamflow.

Locate septic systems an adequate distance from streams.

Pathogens and excess nutrients from on-site sewage disposal can pollute streams. The minimum distance needed between a septic system and a river depends on how diluted the waste matter is, the capacity of the soil to absorb and renovate infiltration, and how long it takes the effluent to travel through the soil.

Control stormwater runoff.

Peak rates of runoff should be controlled to avoid excessive surges in flow. Runoff should be treated to minimize pollution, using such means as direction of sheetflow over vegetated areas, sediment basins, catch-basin sumps, and oil traps.

Control soil erosion.

Controlling erosion reduces the amount of suspended solids, making water clearer. It also reduces deposits of sediment that can bury the channel bed.

Manage instream flows.

Manage release rates from dams, reservoirs, and diversions so that they meet not only human needs, but also the needs of the habitat as water supply and demand vary from season to season.

Provide fish passage.

Whenever possible, dams, bridges, culverts, and channels should include means for fish to move past the structures.

A fish ladder allows fish to migrate past a relatively low dam.

TABLE 98 FISH PASSAGE AT SMALL DAMS
• Remove abandoned dams
• Partially breach dams
• Install bypass channels around dams
• Provide fish ladders

TABLE 99 BENEFITS OF SMALL DAM REMOVAL
• Provides fish passage
• Improves river boating
• Reduces floodwater elevations
• Improves aeration, circulation

- Concentrate low flows in a V-shaped bottom when using a box culvert, or in a single pipe where multiple-pipe structures are used.
- Avoid vertical drops at culvert outlets.
- Avoid sharp-edged riprap.
- Preserve adjacent vegetative cover and sheltered areas in the channel.
- Provide pools at both ends of the culvert.

Fish Passage at Dams

The movement of fish past dams is complicated by the large numbers of existing structures and their height above the streambed. In some cases, stream restoration may include removal of old dams that are no longer functional, or at least their partial breach. This is viable at many New England run of the river dams that were formerly used by textile mills and industries that are inactive or abandoned.

Fish ladders are being built at some key dams, especially where they block the restoration of runs of anadromous fish. However, fish ladders are expensive and their effectiveness is sometimes limited. Fish ladders are being used at low-head hydroelectric generating dams that are located on small and mid-size rivers. The photo at left shows such a fish ladder.

Many dams have sediment deposits in their upstream impoundments and this becomes a key consideration in removing or breaching dams. In some cases, the sediments may include contaminants that should not be allowed to wash downstream. In other cases, sediments could erode and block downstream channels, or bury the substrate.

Several options are available for sediment deposits at dam removal sites. They include full sediment removal, dredging a new channel through the sediments and stabilizing those remaining, or allowing sediments to be scoured until reaching equilibrium conditions.

Dealing With Unavoidable Impacts: The Mitigation Process

Where it is determined that human activities will have an unavoidable impact on a river system, it is suggested — and often required — that the project sponsor mitigate the damage. A suggested process to accomplish that is shown in Figure 76.

TABLE 100

INSTREAM HABITAT MITIGATION MEASURES

The detrimental effects of channelization can be reduced by providing instream mitigation measures compatible with the natural stream and floodplain morphology. Specific stream corridor methods of mitigating environmental damage include:

Deflectors
Flow deflectors made of rock or timber are constructed on the streambed adjacent to the banks to concentrate low flows. They can be quite effective, but they need to be low enough so that they do not obstruct floodflows. They may be arranged so as to encourage the flow to meander. Deflectors should meet the bank at a 30-degree angle, and have a sturdy fill of stone between the front face and the bank to prevent erosion.

Check dams and sills
Small check dams, frequently only one to three feet high, can be installed across steeper streams to form low-velocity pools. Scour holes form downstream. Aeration of the water occurs at the overflow point. Check dams may be made of timber, buried logs, rocks, or concrete. They should have a depressed center, to concentrate low flows. Check dams should not be used where they may create large pools that drown out riffles. The downstream banks should be lined for a length three times the channel width to prevent them from being undermined by expansion of the scour hole.

Boulders
Rocks of various sizes can be placed in the stream to provide shelter and create irregular low-flow patterns. Their height should be greater than the normal flow depth. They can be especially effective in deep streams where deflectors could be damaged by floodwaters. Since scour can occur around boulders, they should not be placed next to the banks or placed with the longer axis perpendicular to flow. A boulder should be large enough to resist movement, but should not obstruct more than 20 percent of the channel width. Boulders may be placed in a series to form cascades or at the inside of bends. They may also be placed for aesthetic purposes.

Pools
The channel should have well-defined pools, three feet or deeper, in areas where they will tend to be self-cleaning. They may be at bends, or at scour holes below check dams. They should have a length greater than the stream width and be separated from each other by a distance equal to three channel widths.

Riffles
Areas of shallow, fast flows should be created between pools. The goal is to create areas of fast flows over gravel beds. They may be formed with cobble beds or rock sills and should be located between bends.

Stream velocity
Variable stream velocities encourage a diverse habitat. In cold-water areas, velocities of two feet per second are desirable for trout and other game fish. They remove fine sediments, thereby creating a coarse-grained bed.

Wetlands
It is desirable to preserve or even create swamps and marshes alongside new channels. They provide food sources and help mitigate the damages caused by any reduction in original channel length.

Bed material
In rivers not subject to deposition, a coarse bed of gravel and cobbles can be provided, if not already present. This type of bed provides the best conditions for cold-water fisheries and is least susceptible to movement. Maintaining it requires adequate flow concentration and velocity to prevent siltation.

Vehicle crossing
These should be confined to bridges or improved fords with a riprap base to minimize erosion and disturbance of streambank vegetation. Access should be limited to curtail recreational vehicles.

Vegetation
Every effort should be made to preserve existing vegetation and to provide new vegetation along the banks. This provides cover and food sources for wildlife, filters runoff, and shades the water, as well as supplying organic material. During river projects, existing vegetation may be preserved by working from the streambed or by working on only one side of the river. Species of new vegetation should be carefully chosen in order to provide sources of food during all seasons and minimize flow obstructions.

Buffer zone
Vegetated buffers are desirable on both sides of the channel to shelter the river from the effects of intense land use, provide wildlife cover and trap eroded soils before they enter the river. The width should be based on soil, slope and vegetation conditions.

Proper placement of spoils
Excavated soils should not be used to form large fill areas on the floodplain. Some states allow limited local spoil piles to make for topographical diversity. But they should not be placed so that they form continuous levees blocking local drainage or floodways.

Minimizing erosion of disturbed soils
Soils disturbed by the work should be protected from erosion as soon as possible to reduce sedimentation, which could bury valuable food sources and spawning beds.

Avoiding migratory barriers
Eliminate or minimize barriers such as culverts, bridges and dams that interfere with the movement of fish, and provide a means of passage around barriers whose placement cannot be avoided.

Aesthetics
Channel projects should be designed with an eye toward aesthetics. Disturbed areas should be well landscaped, vistas should be provided, and litter removed.

Proper schedule
Consult with fisheries officials so that construction is avoided during spawning seasons and migration runs.

FIGURE 76

MITIGATING UNAVOIDABLE IMPACTS: A SUGGESTED PROCESS

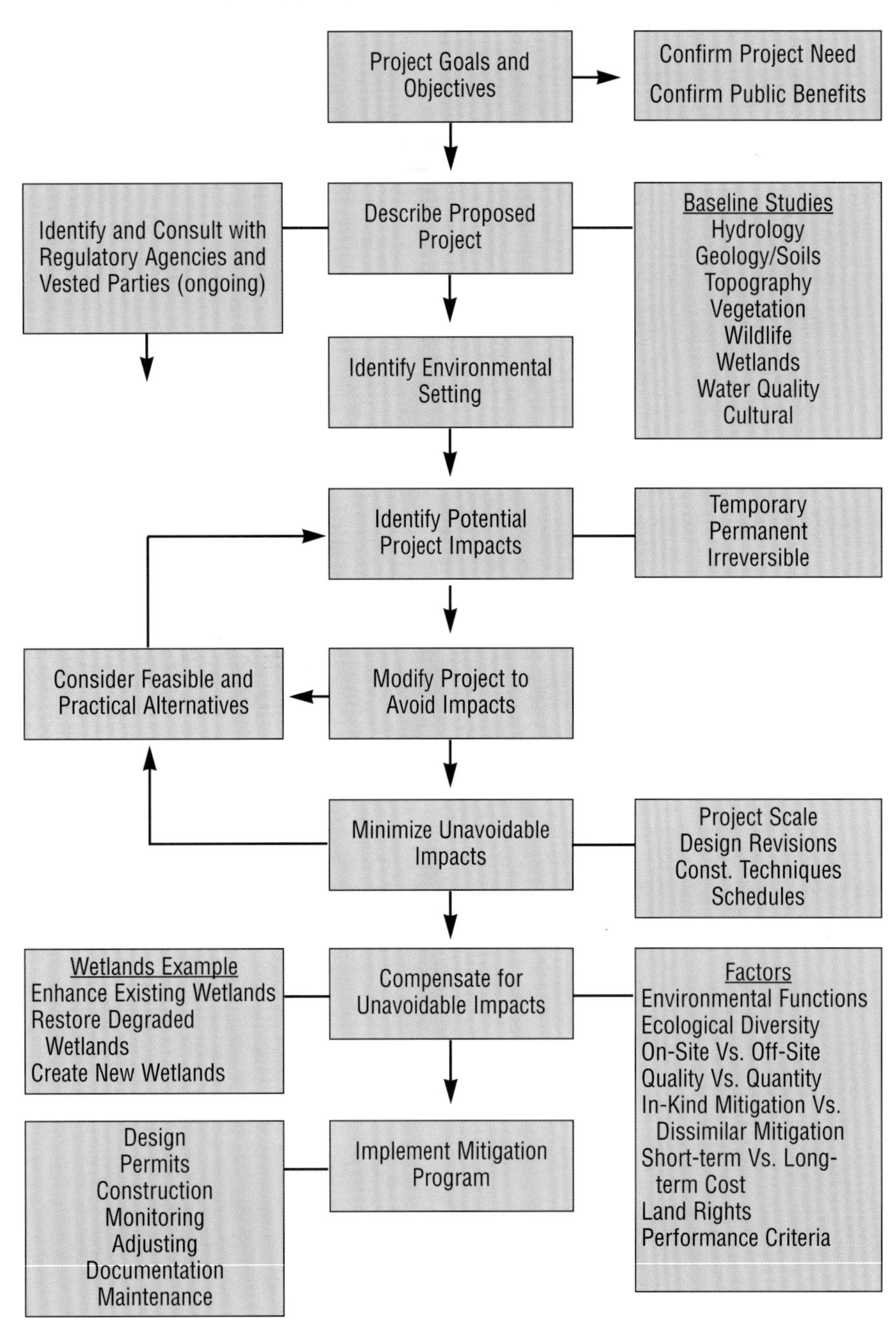

Appendix

Government Programs

In a democracy, citizens must participate in public processes to ensure the long-term protection of their basic rights. So, too, with rivers. Government at all levels has a big impact on what happens to rivers, and citizen participation is critical to the protection of rivers. Those wishing to conserve river resources must participate in public processes, such as public hearings, budget adoption, and evaluation of proposed construction activities.

Government may regulate what goes in or comes out of a river, purchase open space, manage land and water resources, or impact waters through construction and building activities. In many cases, government is the arbitrator between conflicting demands, such as that between maintaining enough water for fisheries and diversions of water for water supply, irrigation or industrial purposes.

Local governments in Connecticut, and many other states, have agencies or commissions that:

- Regulate the use of wetlands.
- Regulate development within floodplains.
- Control erosion and sedimentation during construction.
- Purchase and maintain open space.
- Build and maintain flood-protection works.
- Clear clogged or blocked channels.
- Regulate land uses through zoning ordinances.
- Plan for future development.
- Provide disaster response from such calamities as hurricanes, floods or dam failures.
- Develop and provide water-based recreation programs.

Citizens can check with their local community about the specific regulations and forms of assistance. Connecticut towns have inland wetland commissions, zoning boards, conservation commissions, and mayors or first selectmen whose offices can answer most questions concerning rivers.

State governments, such as that of Connecticut, often have many of the same responsibilities, plus additional ones. They obtain these duties from legislative authority. Among the common state responsibilities are:

- Designation of a single state agency to protect the state's natural resources.
- Regulation of fresh water or inland waters and coastal or tidal wetlands.
- Regulation of state actions that affect rivers, such as constrution of buildings and roads.
- Alteration or modification of flow.
- Oversight of certain municipal functions.
- Control over dredging activities.
- Control over the placement of structures within waterways.
- Establishing and regulating flood zones.
- Controlling pollution of surface waters and groundwater.
- Stocking rivers, lakes and ponds with fish.
- Controlling and licensing hunting and fishing.
- Providing for dam inspections, repair and construction.
- Planning, building and maintaining flood-control structures.
- Purchasing open space.
- Regulating diversions of water from one river basin to another.
- Requiring minimum flows in certain rivers.

- Coordinating river programs.
- Establishing greenways and river belt protection zones.
- Licensing and/or overseeing drinking-water and wastewater treatment facilities.
- Responding to spills into waterways.
- Controlling public trust lands, which are certain lands beneath navigable waterways or certain tide lines.
- Providing disaster response from such calamities as hurricanes, floods or dam failures.
- Protection of drinking water.
- Licensing of boaters.
- Providing guidance for lake management.
- Providing guidance for inland wetland agencies on wetland and watercourse management.

Most of Connecticut's river-related activities are managed or regulated by the Connecticut Department of Environmental Protection, 79 Elm Street, Hartford, CT 06106. Please consult the government section of your phone book for the numbers of appropriate agencies. The general number for the Department of Environmental Protection at the time of this book's publication was 860-424-3000.

Regional and county governments outside of Connecticut often have a mixture of the powers and duties of state and local governments. In Connecticut the regional councils of governments usually have a land-use and development planning presence, and may have other duties specific to their region. A listing of these agencies in Connecticut may be obtained from the bookstore at the Connecticut Department of Environmental Protection.

The federal government has powers designated by Congress, or interpreted by the courts to be within the jurisdiction of the federal government. Some of these powers are exclusive and others are not. Among the federal responsibilities are:

- Determining the need for flood-protection works, and constructing them in cooperation with other government entities.
- Providing disaster response from such calamities as hurricanes, floods or dam failures.
- Setting pollution prevention standards.
- Assisting states and local governments with water pollution prevention measures.
- Providing a national program for flood insurance.
- Regulating alterations of navigable waterways.
- Conducting research on water resources.
- Providing water monitoring programs.
- Protecting drinking water.
- Settling disputes between states.

A number of federal agencies may provide assistance regarding issues related to rivers. Among them are:

- Natural Resource Conservation Service
- U.S. Army Corps of Engineers
- U.S. Geological Survey
- Environmental Protection Agency
- Federal Emergency Management Agency
- National Weather Service
- U.S. Fish and Wildlife Service

To reach a federal agency, consult the government section of your phone book, or call directory assistance and ask for the United States Federal Information Center. When this book was published, that number was 1-800-688-9889.

Glossary

Aggradation: A general raising of a riverbed by deposition of sediment over a long period of time and over a long distance.

Alluvial river: A river that has a bed and banks of unconsolidated sedimentary material subject to erosion, transportation and deposition by the river. The channel geometry and flow conditions are interrelated.

Alluvium: Deposits of clay, silt, sand, gravel, or other particulate material in a streambed, on a floodplain or delta, or at the base of a mountain.

Anadromous fish: Migratory species, such as salmon, shad and striped bass, that are born in fresh water, spend most of their lives in estuary and ocean waters, and return to fresh water to spawn.

Aquifer: A geologic formation that contains sufficient saturated permeable material to yield significant quantities of water to wells and springs.

Armor: A natural layer of particles, usually gravel and cobble sizes, that may cover the surface of the streambed as a coarse residue after erosion of the finer bed materials. The layer, often just one to three particles thick, inhibits the erosion of underlying particles.

Average discharge (surface water): As used by the U.S. Geological Survey, the average recorded discharge of all years on record, whether consecutive or not. The term "average" generally is reserved for average of entire record and "mean" is used for averages of shorter periods, namely, daily, monthly or annual mean discharges.

Backwater: The lowlands adjacent to a stream that become flooded during periods of high water and which are sheltered from the main current.

Bank failure: Collapse or slippage of bank material into a stream channel.

Bar: A sand or gravel deposit found on the bed of a stream that is often exposed only during periods of low water.

Base flow: The sustained dry-weather flow of water in a stream.

Bed: The bottom of a channel.

Bed forms: Generic terms used to describe small irregularities on the bed, including ripples, dunes and antidunes.

Bed load: Sediment that moves by saltation (jumping), rolling, or sliding along and on the riverbed as it is transported downstream.

Bed material: The material composing a streambed.

Bedrock: General term for solid rock that underlies soil or other unconsolidated material.

Bed slope: The inclination of the channel bottom.

Best Management Practices (BMPs): The most environmentally, socially and economically appropriate measures that can be applied in order to accomplish a desired end.

Buffer zone: An area separating two areas of potentially conflicting use. The objective of a buffer zone along a river is to reduce the possibility that land use could have an adverse impact on water quality.

Caving: The collapse of a streambank by undercutting due to erosion of the toe or the soil above the toe.

Channel: A natural or man-made linear waterway that continuously or periodically contains a stream of moving water.

Channelization: The widening, deepening and realignment of natural stream channels in order to increase their flow capacity or to reclaim land.

Channel restoration: The process of improving the flow conditions and ecological value of rivers through minor operations to reduce erosion, remove trash, vegetate areas, etc. It does not include major channelization work.

Check dam: A structure placed bank to bank across a stream to reduce flow velocities.

Chute: See "cut-off."

Clay: Cohesive and generally impervious soil whose individual particles are too small to be visible to the unaided eye.

Consumptive use: Water removed from the immediate aquatic environment through evaporation, transpiration, human consumption, agriculture, industry, etc. Sometimes called water consumed or water depletion.

Cubic feet per second: A common method for measuring the volume of water, in cubic feet, flowing past a point in one second of time. Abbreviated CFS.

Cubic feet per second per square mile: A common method of describing the flow rate on a unit basis, from one square mile of watershed area. Abbreviated CSM.

Cut bank: The outside bank of a channel bend, often eroding and on the opposite side of the stream from a point bar.

Cut-off: A channel cut across the neck of a bend, sometimes called a meander cut-off or chute.

Degradation: A general lowering of the riverbed by erosion over a long period of time and over a long distance.

Dike (groin, spur, jetty, deflector): A structure designed either to reduce water velocity as streamflow passes through it, thus encouraging sedimentation instead of erosion (permeable dike), or to deflect erosive currents away from a streambank (impermeable dike).

Discharge: The volume of water passing through a channel during a given time, usually measured in cubic feet per second.

Dissolved solids: Minerals and organic matter dissolved in water.

Diversion: A turning aside or alteration of the natural course of water flow, usually considered to be a physical departure from the natural channel. Examples include a consumptive use directly from a stream, such as by livestock watering, or the taking of water through a canal, pipe or other conduit.

Drainage basin: The land area drained by a river, also called a watershed.

Drainage divide: Boundary between one drainage basin and another.

Estuary: That part in the lower course or mouth of a river in which the river's current meets sea water.

Eutrophication: Process by which an aging lake or pond becomes enriched with plant nutrients, most commonly phosphorus and nitrogen.

Evaporation: Process by which water is changed from the liquid or solid state to the vapor state.

Evapotranspiration: A collective term that includes water discharged to the atmosphere as a result of evaporation from the soil and surface waters, as well as by plant transpiration.

Extratropical storm: A storm that originated as a tropical storm or hurricane that has moved north and lost the characteristics of a tropical cyclone but which may still produce torrential rains and, potentially, flooding.

Fecal coliform bacteria: Bacteria present in the gut or feces of warmblooded animals; their presence in water is an indicator of possible sewage pollution.

Flood: A high streamflow rate that overtops the channel banks.

Floodplain: The portion of a valley that is periodically inundated by water in excess of the channel capacity. It is often formed of sediment deposits.

Fluvial: Pertaining to a river or stream.

Geomorphology: The study of the shape and form of the Earth's surface.

Glacial drift (glacial till): Rock material (clay, silt, sand, gravel, boulders) transported and deposited by a glacier.

Headcutting: The action of an upstream-moving waterfall or locally steep channel bottom with rapidly flowing water through an otherwise placid stream. These conditions often indicate that a readjustment of a stream's discharge and sediment-load characteristics is taking place.

Herbaceous: Having the characteristics of an herb, a plant with no persistent woody stem above ground.

Humus: The dark organic material on or in the soil, from the decomposition of plant material.

Hydraulics: The branch of engineering that deals with the motion of water and other fluids.

Hydrologic modification: Any alteration of the terrain that changes the movement, distribution, flow, or circulation of surface water or groundwater, such as construction of dams, stream channelization or paving extensive areas of land.

Hydrology: The branch of geology dealing with the distribution, properties and effects of water on or below the surface of the Earth.

Infiltration: The movement of water into soil or porous rock.

In-stream use: Water use taking place within the stream channel for such purposes as hydroelectric power generation, navigation, water-quality improvement, fish propagation, and recreation. Sometimes called nonwithdrawal use or in-channel use.

Intermittent stream: A stream or part of a stream that flows only in direct response to precipitation, receiving little or no water from springs, snowmelt or other sources. It may be dry for a large part of the year, generally more than three months. Its course may be marked by an eroded channel, sediment deposits, scour, or transport of leaf litter or soil.

Lower bank: The portion of a streambank below the elevation of the stream's average water level.

Nonpoint pollution: Pollution from broad areas — such as storm drains, roads, areas of fertilizer and pesticide application, and leaking sewer systems — rather than from discrete points.

Nutrient: Referring to water quality elements, including nitrogen and phosphorus, that stimulate growth of algae and other plants.

Organic matter: Any part of any substance that once had life. The waste, byproduct or decomposed material of animals or plants.

Overbank drainage: Water flow over the top of a bank and down a slope.

Palustrine: Of or related to a marsh, bog or swamp.

Peak flow: The maximum instantaneous rate of flow during a flood.

Perennial: Persisting more than two years. Also a plant that uses the same root system to produce new growth each year.

Perennial stream: A stream that normally has water in its channel at all times.

Phytoplankton: Free-floating microscopic plants in bodies of water.

Point bar: The bar on the inside of a bend of a stream that has built up due to sediment deposition.

Point source of pollution: A discrete source of pollution, such as a pipe or other man-made conveyance of waste.

Pool: A section of a river with relatively deep water and low flow velocities.

Reach: A short section of a stream's length, between specified points.

Regime rivers (also graded rivers): A situation where the rate of erosion, deposition, flow rates, and sediment conditions in an alluvial channel are in dynamic equilibrium.

Revetment: A facing of stone, bags, blocks, pavement, etc., used to protect a bank against erosion.

Riffles: A relatively shallow section of a river with high-velocity, turbulent flow.

Rill erosion: Erosion caused by sheetflow moving down a fairly steep slope, cutting small, fairly evenly spaced channels up to 1 foot deep into the ground.

Riparian vegetation: Vegetative growth along the banks of a stream.

Riprap: A man-made lining of loose rock on the bed or banks of a channel to prevent erosion.

River: A large stream of water from a drainage basin of significant size, normally confined in a natural watercourse or channel.

Rivulet: A very small stream, also called a streamlet.

Runoff: That part of the precipitation that appears in surface-water bodies. It is the same as streamflow unaffected by artificial diversions, storage or other human works in or on the stream channels.

Saline water: Water generally considered unsuitable for human consumption or for irrigation because of its high content of dissolved solids. The concentration of dissolved solids is generally expressed in milligrams per liter, with 35,000 mg/L defined as sea water.

Scour: Excessive erosion of a streambed or bank, usually resulting from concentrated flows or extremely high flow rates or velocities, or from disruption of a natural channel.

Sediment: Fragmented and decomposed rock and soils that are transported and deposited by water.

Sediment discharge (or sediment load): The quantity of sediment that is carried past any cross section of a stream in a unit of time.

Sediment yield: The total sediment outflow from a watershed or a drainage area at a point of reference and in a specified period of time. This is equal to the sediment discharge from the drainage area.

Seep: A natural discharge of groundwater emanating from a relatively large, poorly defined area of soil or rock.

Sheet erosion: Removal of a thin, fairly uniform layer of soil from the land surface as a result of sheetflow.

Sheetflow: Runoff that flows uniformly like a sheet over the ground.

Shrub: A woody plant that at maturity is usually less than 20 feet tall and generally exhibits several erect, spreading, or prostrate stems and has a bushy appearance.

Snagging: Removal or clearing of logs, trees, tree stumps, or other debris from a stream channel.

Spring: A natural discharge point where groundwater issues from soil or rocks in concentrated flow.

Stratified drift: Glacial sediments sorted according to size by the glacial meltwaters. Deposits of stratified drift have a high capacity to store water between particles and are major sources of drinking water.

Stream: A flowing body of water moving in a particular direction in a channel.

Streambank erosion: Removal of soil particles from a bank slope primarily due to water action. Climatic conditions, ice and debris, chemical reactions, and changes in land and stream use may also lead to bank erosion.

Suspended load: The portion of the sediment load that is transported by water with little contact with the streambed. It includes suspended bed material and the wash load.

Suspended sediment: Sediment that is transported by a stream: both colloidal material, and larger, fragmental material, mineral and organic, that is maintained in suspension by the upward components of turbulence and currents.

Swale: A small channel or ditch, usually vegetated, with intermittent flow.

Terrace: A former floodplain no longer subject to frequent flooding, generally caused when a river channel erodes a deeper channel below the floodplain elevation.

Thalweg: An imaginary line along the length of a river connecting the deepest points.

Tidal prism: The total volume of water that flows in and out of a tidal river or estuary during each tide cycle; the difference in volume between high tide and low tide.

Tidal river: A river in which the flow's rate, direction or elevation is affected by tides.

Till: See "glacial drift."

Toe: The break in slope at the foot of a streambank where it meets the streambed.

Top bank: The break in slope between the bank and the surrounding terrain.

Transpiration: The release of water vapor by plants during photosynthesis.

Tree: A woody plant that at maturity is usually 20 feet or more in height and generally has a single trunk unbranched to at least 3 feet above the ground, with a more or less definite crown.

Turbidity: The opaqueness or reduced clarity of a fluid due to the presence of suspended matter.

Turbulent flow: A state of flow in which the water is agitated by cross currents and eddies, and in which flow velocities vary irregularly in magnitude and direction.

Uniform flow: A flow that is constant in both velocity and direction along the stream lines.

Upland: Land distinct from the riparian or wetland zone and above it in the watershed.

Upper bank: The portion of a streambank above the elevation of the stream's average water level.

Wash load: That portion of the sediment load that is smaller than the bed material and is being transported from upstream areas. The quantity transported depends on the rate of supply, and usually consists of clay and silt that remains in suspension. It may include sand if the bed material is gravel.

Wastewater: Water that contains dissolved or suspended solids as a result of human use.

Watercourse: A stream of water flowing in a channel with well-defined banks.

Watershed: An area confined by drainage divides, usually having only one streamflow outlet.

Water table: The upper surface of the unconfigured zone of saturation in the soil.

Wave attack: Impact of waves on a streambank.

Zooplankton: Free-floating microscopic animals in bodies of water.

ibliography

CHAPTER ONE

Handman, Elinor H., Haeni, F. Peter, and Thomas, Mendall P., *Water Resources Inventory of Connecticut, Part 9, Farmington River Basin*. Connecticut Water Resources Bulletin No. 29, U.S. Geological Survey in cooperation with Connecticut Department of Environmental Protection. (Hartford, Conn., 1986)

Hunter, Bruce, and Meade, Daniel, *Precipitation in Connecticut*. Connecticut Department of Environmental Protection Bulletin No. 6. (Hartford, Conn., 1983)

Leopold, Luna B., *Hydrology for Urban Land Planning: A Guidebook on the Hydrologic Effects of Urban Land Use*. U.S. Geological Survey Circular No. 554. (Washington, D.C., 1968)

Miller, David, and Jones, Bryon, *Precipitation*. University of Connecticut Agricultural Climate Station Bulletin 447. (Storrs, Conn., 1978)

Miller, David, Warner, Glenn S., DeGaetano, Arthur, *Rainfall in Connecticut*. University of Connecticut, Department of Natural Resources. (Storrs, Conn., 1997)

Milone & MacBroom, Inc., *Reservoir Yield Manual*. Report to Connecticut Department of Health. (Hartford, Conn., 1989)

Moody, David W., Chase, Edith B., and Aronson, David A., *National Water Summary 1985 – Hydrologic Events and Surface-Water Resources*. U.S. Geological Survey Water Supply Paper 2300. (Washington, D.C., 1986)

Sauer, V.B., Thomas, W.O. Jr., Stricker, V.A., and Wilson, K.V., *Flood Characteristics of Urban Watersheds in the United States*. U.S. Geological Survey Water Supply Paper 2207. (Washington, D.C., 1983)

University of Connecticut, *A Connecticut Water Primer*. University of Connecticut Cooperative Extension Service bulletin. (Storrs, Conn., 1975)

Weiss, L.A., *Floodflow Formulas for Urbanized and Non-Urbanized Areas of Connecticut*. U.S. Geological Survey bulletin. (Hartford, Conn., 1975)

CHAPTER TWO

Adams, Sherman, *The Maritime History of Wethersfield, Vol. I.* Wethersfield Historical Society. (Wethersfield, Conn., 1977)

Barber, John Warner, *Connecticut Historical Collections*. B.L. Hamlen. (New Haven, Conn., 1836)

Blench, Thomas, *Scour at Bridge Waterways*. Transactions of the American Society of Civil Engineers, Vol. 127. (1962)

Blench, Thomas, *Mobile-Bed Fluviology*. University of Alberta Press (Edmonton, Canada, 1969)

Bray, Dale I., *Estimating Average Velocity in Gravel Bed Rivers*. Journal of the Hydraulics Division of the American Society of Civil Engineers, September 1979. (New York)

Bray, Dale I., and Church, Michael, *Technical Notes*. Proceedings paper 15791, Journal of the Hydraulics Division of the American Society of Civil Engineers. (New York, 1973)

Brookes, Andrew, *Channelized Rivers: Perspectives for Environmental Management*. John Wiley & Sons. (Chichester, England, 1988)

Chang, Howard H., *Fluvial Processes in River Engineering*. John Wiley & Sons (New York, 1988)

Chang, Howard H., *Geometry of Rivers in Regime*. Journal of the Hydraulics Division of the American Society of Civil Engineers, Vol. 105, No. HY 6. (New York, 1979)

Chang, Howard H., *River Changes: Adjustments of Equilibrium*. American Society of Civil Engineers Journal of Hydraulic Engineering, Vol. 112, No. 1. (New York, 1986)

Culbertson, D.M., Young, L.E., and Brice, J.C., *Scour and Fill in Alluvial Channels*, U.S. Geological Survey Open File Report 69-1967.

Dreyer, Glenn D., and Niering, William A., *Tidal marshes of Long Island Sound: Ecology, History and Restoration*. Connecticut College Arboretum, Bulletin No. 34. (New London, Conn., 1995)

Dury, G.H., *Principles of Underfit Streams*. U.S. Geological Survey Professional Paper 452-A. (Washington, D.C., 1964)

Flint, Richard Foster, *The Glacial Geology of Connecticut*. State Geological and Natural History of Connecticut, Bulletin 47. (Hartford, Conn., 1936)

Fox, Helen A., *The Urbanizing River: A Case Study in the Maryland Piedmont*. In *Geomorphology and Engineering*, Donald R. Coates, editor. Dowdon, Hutchinson & Ross. (Stroudsburg, Pa., 1976)

Henderson, Francis M., *Stability of Alluvial Channels*. Journal of the Hydraulics Division of the American Society of Civil Engineers, November 1961. (New York)

Hydraulic Engineering Center of the U.S. Army Corps of Engineers, *Sediment Transport, Hydrologic Engineering Methods for Water Resources Development, Vol. 12*. (Davis, Calif., 1977)

Keller, E.A., *Bed Material Sorting in Pools and Riffles*. Discussion in Journal of the Hydaulics Division of the American Society of Civil Engineers, September 1983. (New York)

Keller, E.A., *Development of Alluvial Stream Channels, A Five Stage Development*. Geological Society of America Bulletin, Vol. 83, May 1972.

Keller, E.A., and Melhorn, W.M., *Rhythmic Spacing and Origin of Pools and Riffles*. Geological Society of America Bulletin, Vol. 89, May 1978.

Kellerhals, Rolf, Church, Michael, and Bray, Dale, *Classification and Analysis of River Processes*. Journal of the Hydraulics Division of the American Society of Civil Engineers, July 1976. (New York)

Kulp, Kennith, *Suspended Sediment Characteristics of Muddy Brook at Woodstock Connecticut*. Connecticut Water Resources Bulletin No. 43, U.S. Geological Survey. (Hartford, Conn., 1991)

Leopold, L.B., *River Floodplains: Some Observations on Their Formation*. U.S. Geological Survey Professional Paper 282-G. (Washington, D.C., 1957)

Leopold, L.B., and Maddock, T., *The Hydraulic Geometry of Stream Channels and Some Physiographic Implications*. U.S. Geological Survey Paper 252. (Washington, D.C., 1953)

Leopold, L.B., and Wolman, M.G., *River Channel Patterns: Braided, Meandering and Straight*. U.S. Geological Survey Professional Paper 282-B. (Washington, D.C., 1953)

Leopold, L.B., Wolman, M.G., and Miller, J.P., *Fluvial Processes in Geomorphology*. W.H. Freeman. (San Francisco, Calif., 1964)

MacBroom, James Grant, *Applied Fluvial Geomorphology*. University of Connecticut, Institute of Water Resources bulletin, March 1981. (Storrs, Conn.)

Simons, Daryl B., and Albertson, M.L., *Uniform Water Conveyance Channels in Alluvial Material.* Transactions of the American Society of Civil Engineers. (New York, 1963)

Simons, Daryl B., and Sentuck, Fuat, *Sediment Transport Technology*, Water Resources Publications. (Fort Collins, Colo., 1976)

Thomas, Mendall, *Depth of 100 Year Flood Versus Drainage Area.* U.S. Geological Survey. (Hartford, Conn.)

Vanoni, Vito, editor, *Sedimentation Manual*, American Society of Civil Engineers, Manual of Practice No. 54. (New York, 1977)

Wolman, M.G., and Leopold, L.B., *River Flood Plains: Some Observations on Their Formation*, U.S. Geological Survey Professional Paper 282-C.

Woodyer, K.D., *Bankfull Frequency in Rivers.* Journal of Hydrology, April 1968.

Wullard, John C., *Where Is the Connecticut River?* Article in Wethersfield (Conn.) Post, July 20, 1972.

CHAPTER THREE

Adams, L.W., and Dove, L.E., *Wildlife Reserves and Corridors in the Urban Environment.* National Institute for Urban Wildlife. (Columbia, Md., 1989)

American Society of Civil Engineers Task Committee on Sediment Transport and Aquatic Habitats, *Sediment and Aquatic Habitat in River Systems.* Journal of the Hydraulics Division of the American Society of Civil Engineers, May 1992. (New York)

Benyus, Janine M., *The Field Guide to Wildlife Habitats of the Eastern United States.* Fireside. (New York, 1989)

Bloom, Arthur, and Ellis, Charles, *Post Glacial Stratigraphy and Morphology of Coastal Connecticut.* State Geological and Natural History of Connecticut Guidebook No. 1. (Hartford, Conn., 1965)

Cowardin, L.M., et al., *Classification of Wetlands and Deepwater Habitats of the United States.* U.S. Fish and Wildlife Service. (Washington, D.C., 1979)

DeGraaf, Richard M., and Rudis, Deborah D., *New England Wildlife: Habitat, Natural History, and Distribution.* U.S. Forest Service Northeastern Forest Experiment Station General Technical Report NE-108. (Broomall, Pa., 1986)

Dowhan, J.J., and Craig, R.J., *Rare and Endangered Species of Connecticut and Their Habitats.* State Geological and Natural History Survey of Connecticut, Department of Environmental Protection Natural Resources Center. (Hartford, Conn., 1976)

Fedler, A.J., and Nickum, D.M., *The 1991 Economic Impact of Sport Fishing in Connecticut.* Sport Fishing Institute. (Washington, D.C., 1993)

Gore, James A., editor, *The Restoration of Rivers and Streams: Theories and Experience.* Butterworth Publishers. (Boston, 1985)

Greeson, P.E., Clark, J.R., and Clark, J.E., *Wetland Functions and Values: The State of Our Understanding.* American Water Resources Association. (Minneapolis, 1978)

Golet, F.C., and Larson, J.S., *Classification of Freshwater Wetlands in the Glaciated Northeast.* U.S. Fish and Wildlife Service Resource Publication 116. (1975)

Magee, Dennis W., *Freshwater Wetlands, A Guide to Common Indicator Plants of the Northeast.* University of Massachusetts Press. (Amherst, Mass., 1981)

Maryell, Richard, *The Potential Impacts of Clearing and Snagging on Stream Ecosystems.* U.S. Fish and Wildlife Service. (Washington, D.C., 1978)

Metzler, K.J., and Tiner, R.W., *Wetlands of Connecticut*, State Geological and Natural History Survey of Connecticut, Report of Investigations No. 13. (1992)

Miller, Jack G., and Tibbott, Ron, *Fish Habitat Improvement for Streams.* Pennsylvania Fish Commission. (Harrisburg, Pa.)

National Institute for Urban Wildlife, *Wildlife Reserves and Corridors in the Urban Environment: A Guide to Ecological Landscape Planning and Resource Conservation.* (Columbia, Md.)

Niering, William A., *Wetlands.* Chanticleer Press. (New York, 1985)

Odum, William E., Smith, Thomas J. III, Hoover, John K., and McIvor, Carole C., *The Ecology of Tidal Freshwater Marshes of the United States East Coast: A Community Profile.* University of Virginia Department of Environmental Sciences. (Charlottesville, Va., 1984)

Odum, William E., et al., *The Ecology of Tidal Freshwater Marshes of the United States East Coast: A Community Profile.* U.S. Fish and Wildlife Service. (Washington, D.C., 1984)

Platt, Rutherford H., editor, *Regional Management of Metropolitan Floodplains.* Institute of Behavioral Sciences, University of Colorado. (Boulder, Colo., 1987)

Reid, George K., *Ecology of Inland Waters and Estuaries.* Van Nostrand Reinhold Co. (New York, 1961)

Simpson, Philip, et al., *Manual of Stream Channelization Impacts on Fish and Wildlife.* U.S. Fish and Wildlife Service. (Washington, D.C., 1982)

Tiner, Ralph, *Wetlands of the United States, Current Status and Recent Trends.* U.S. Fish and Wildlife Service. (Newton, Mass., 1984)

Thomas, Carl H., Robbins, Samuel W., and McCandless, Donald E., *Fish and Wildlife Considerations in Small Reservoir and Channel Construction in the Northeast.* American Society of Civil Engineers, Annual Environmental Engineering Convention, Meeting Preprint 2388. (Kansas City, Mo., 1974)

Thomson, Keith Stewart, *Saltwater Fish of Connecticut.* State Geological and Natural History Survey of Connecticut, Bulletin 105. (Hartford, Conn., 1978)

U.S. Army Corps of Engineers, New England Division, *Buffer Strips for Riparian Zone Management.* (Waltham, Mass., 1991)

U.S. Fish and Wildlife Service, *Western Reservoir and Stream Habitat Improvements Handbook.* (Washington, D.C., 1978)

U.S. Fish and Wildlife Service and U.S. Department of Commerce, 1993 Bureau of the Census. *1991 National Survey of Fishing, Hunting, and Wildlife-Associated Recreation.*

Vanote, Robin, et al., *The River Continuum Concept.* Government of Canada, Canadian Journal of Fisheries and Aquatic Sciences, Vol. 37, No. 1. (1980)

Whitworth, Keith Steward, *Saltwater Fisheries of Connecticut.* State Geological and Natural History Survey of Connecticut, Bulletin 101. (Hartford, Conn., 1968)

Wisconsin Department of Natural Resources, *Guidelines for Management of Trout Stream Habitat in Wisconsin.* Technical Bulletin No. 39. (Madison, Wis., 1967)

CHAPTER FOUR

Applied Biology Inc., *Coastal Marinas Assessment Handbook*, U.S. Environmental Protection Agency. (Washington, D.C., 1985)

Bertness, Mark D., *The Ecology of a New England Salt Marsh.* American Scientist, May 1992. (New York)

Bongiorno, Salvatore, et al., *A Study of Restoration in Pine Creek Salt Marsh, Fairfield, Connecticut.* Fairfield University. (Fairfield, Conn., 1983)

Canter, Larry, *Environmental Impact of Water Resources Projects.* Lewis Publishers. (Chelsea, Mich., 1985)

Daiber, Franklin, *Conservation of Tidal Marshes*, Von Nostrand Reinhold Co. (New York, 1986)

Davis, Richard, Jr., editor, *Coastal Sedimentary Environments*, Springer-Verlag. (New York, 1985)

Dyer, Keith R., *Coastal and Estuarine Sediment Dynamics.* John Wiley & Sons. (New York, 1986)

Jenkins, Joseph Daniel, and Johnson, Harold Malcolm, *Flood Profiles in Combined Tidal-Freshwater Zones.* Journal of the Hydraulics Division of the American Society of Civil Engineers, June 1978. (New York)

Kuo, Chin, editor, *Urban Stormwater Management in Coastal Areas*. American Society of Civil Engineers conference report. (New York, 1980)

Matthiessen, John, *Planning for Sea Level Rise in Southern New England*. Sounds Conservancy Inc. (Essex, Conn., 1989)

Meridith, William H. et al. *Guidelines for Open Marsh Management.* Delaware Department of Natural Resources and Environmental Control, Mosquito Control Section. (1985)

Milone & MacBroom, Inc., *Ragged Rock Creek Salt Marsh Restoration Study.* Report to Connecticut Department of Transportation. (1990)

Milone & MacBroom, Inc., *Sybil Creek Salt Marsh Study: Branford, Connecticut.* Report to Connecticut Department of Environmental Protection. (1987)

Meisler, Harold, Leahy, P. Patrick, and Knobel, Leroy L., *Effect of Eustatic Sea-Level Changes on Saltwater-Freshwater Relations in the Northern Atlantic Coastal Plain.* U.S. Geological Survey Water Supply Paper 2255. (Washington, D.C., 1985)

Myrick, Robert M., and Leopold, Luna B., *Hydraulic Geometry of a Small Tidal Estuary: Physiographic and Hydraulic Studies of Rivers.* U.S. Geological Survey Professional Paper 422-B. (Washington, D.C., 1963)

Niering, William A., and Warren, R. Scott, *Vegetation Patterns and Processes in New England Salt Marshes.* BioScience, May 1980.

Odum, William, Smith, Thomas J. III, Hoover, John K., and McIvor, Carole C., *The Ecology of Tidal Freshwater Marshes of the United States East Coast: A Community Profile*. University of Virginia Department of Environmental Sciences. (Charlottesville, Va., 1984)

Patton, Peter C., and Kent, James M., *The Ecology of a New England Salt Marsh.* American Scientist, May 1992. (New York)

Redfield, Alfred C., *Development of a New England Salt Marsh.* Ecological Monographs. (1972)

Reid, George K., *Ecology of Inland Waters and Estuaries*. Van Nostrand Reinhold Co. (New York, 1961)

Roesner, Larry A., et al., *Design of Urban Runoff Quality Controls.* American Society of Civil Engineers. (New York, 1988)

Roman, Charles True, *Tidal Restriction, Its Impact on the Vegetation of Six Connecticut Coastal Marshes.* Master's thesis, Connecticut College. (New London, Conn., 1978)

Roman, Charles True, Niering, William A., and Warren, R. Scott, *Salt Marsh Vegetation Change in Response to Tidal Restriction.* Environmental Management. (1984)

Teal, John M., *The Ecology of Regularly Flooded Salt Marshes of New England.* U.S. Fish and Wildlife Service. (Washington, D.C., 1986)

Tiner, Ralph W., *Wetlands of the United States, Current Status and Recent Trends.* U.S. Fish and Wildlife Service. (Newton, Mass., 1984)

U.S. Department of Commerce, National Technical Information Service, *Citations From Selected Water Resources Abstracts.* (Washington, D.C.)

U.S. Environmental Protection Agency, *The Lake and Reservoir Restoration Guidance Manual.* (Washington, D.C., 1988)

U.S. Environmental Protection Agency, *Projecting Future Sea Level Rise.* (Washington, D.C., 1983)

U.S. Environmental Protection Agency, *Rapid Bioassessment Protocols for Use in Streams and Rivers – Benthic Invertebrates and Fish.* (Washington, D.C., 1989)

Woodhouse, W.W., *Building Salt Marshes Along the Coasts of the Continental United States.* U.S. Army Corps of Engineers Coastal Engineering Research Center, Special Report No. 4. (Fort Belvoir, Va., 1979)

CHAPTER FIVE

Bradley, Joseph N., *Hydraulics of Bridge Waterways.* U.S. Department of Transportation, Federal Highway Administration. (Washington, D.C., 1973)

Brookes, Andrew, *Channelized Rivers: Perspectives for Environmental Management.* John Wiley & Sons. (Chichester, England, 1988)

Chang, Howard H., *Fluvial Processes in River Engineering.* John Wiley & Sons. (New York, 1988)

Clark, John R., *Coastal Ecosystem Management, A Technical Manual.* Robert Kriger Co. (Malabar, Fla., 1983)

Gore, James A., editor, *The Restoration of Rivers and Streams: Theories and Experience.* Butterworth Publishers. (Boston, 1985)

Hammer, T., *Stream Channel Enlargement Due to Urbanization.* Water Resources, Vol. 8, No. 6. (1972)

Keller, Edward A., *Channelization: Environmental, Geomorphic, and Engineering Aspects.* In *Geomorphology and Engineering*, Donald R. Coates, editor. Dowden, Hutchinson & Ross. (Stroudsburg, Pa., 1976)

Keller E.A., and Hoffman, E.A., *Urban Streams, Plight or Amenity.* Journal of Soil and Water Conservation, October 1977. (Ames, Iowa)

Klein, Richard D., *Urbanization and Stream Quality Impairment.* American Water Resources Association, Water Resources Bulletin, August 1979.

Klengeman, Peter, *Hydrologic Evaluations in Bridge Pier Scour Design.* Journal of the Hydraulics Division of the American Society of Civil Engineers, December 1973.

LaGasse, Peter F., et al., *Impact of Gravel Mining on River System Stability.* Journal of the Waterway, Port, Coastal and Ocean Division of the American Society of Civil Engineers, August 1980. (New York)

Leopold, Luna B., *Hydrology for Urban Land Planning: A Guidebook on the Hydrologic Effects of Urban Land Use.* U.S. Geological Survey Circular 554. (Washington, D.C., 1968)

Maryell, Richard, *The Potential Impacts of Clearing and Snagging on Stream Ecosystems.* U.S. Fish and Wildlife Service. (Washington, D.C., 1978)

New Jersey Department of Environmental Protection, Division of Water Resources, Special Report 38. (Trenton, N.J., 1974)

New York State Department of Environmental Conservation, Bureau of Water Quality, *Stream Corridor Management: A Basic Reference Manual.* (Albany, N.Y., 1982)

Patton, Peter, *The Effects of Urbanization on the Hydrology and Geomorphology of Small Watersheds.* Midstate Regional Planning Agency. (Middletown, Conn., 1979)

Robinson, A.M., *The Effects of Urbanization on Stream Channel Morphology.* Conference proceedings presented at Urban Hydrology, Hydraulics and Sediment Symposium, University of Kentucky. (Lexington, Ky., 1976)

Sauer, V.B., Thomas, W.O. Jr., Stricker, V.A., and Wilson, K.V., *Flood Characteristics of Urban Watersheds in the United States.* U.S. Geological Survey Water Supply Paper 2207. (Washington, D.C., 1983)

Schueler, Thomas R., *Controlling Urban Runoff: A Practical Manual for Planning and Designing Best Management Practices.* Metropolitan Washington Council of

Governments, Department of Environmental Programs. (Washington, D.C., 1987)

Schueler, Thomas R., *Mitigating the Adverse Impacts of Urbanization on Streams: A Comprehensive Strategy for Local Government.* Metropolitan Washington Council of Governments. (Washington, D.C., 1991)

Shields, Fletcher Douglas, and Nunnally, Nelson R., *Environmental Aspects of Clearing and Snagging.* American Society of Civil Engineers Journal of Environmental Engineering, February 1984. (New York)

Simons, Daryl B., and Li, Ruh-Ming, *Failure Probability of Riprap Structures.* American Society of Civil Engineers Convention, Meeting Preprint 3752. (Atlanta, 1979)

Simons, Daryl, and Sentuck, Faut, *Sediment Transport Technology*. Water Resources Publications. (Fort Collins, Colo., 1976)

Simpson, Philip W., et al., *Manual of Stream Channelization Impacts on Fish and Wildlife.* U.S. Fish and Wildlife Service. (Washington, D.C., 1982)

U.S. Army Corps of Engineers, *Hydraulic Design of Flood Control Channels.* Engineering Manual EM 1110-2-1601. (Washington, D.C.)

U.S. Army Corps of Engineers Hydraulic Engineering Center, *Effects of Flood Plain Encroachments on Peak Flow.* (Davis, Calif., 1980)

U.S. Department of Agriculture, *Design of Open Channels.* Soil Conservation Service Technical Release 25. (Washington, D.C., revised 1977)

U.S. Department of Transportation, *Design of Stable Channels With Flexible Linings*. Hydraulic Engineering Circular 15. (Washington, D.C., 1975)

U.S. Department of Transportation, *Highways in the River Environment, Hydraulic and Environmental Design Considerations*. Federal Highway Administration Training and Design Manual. (Washington, D.C., 1975)

U.S. Geological Survey, *Effects of Urbanization on Peak Streamflows in Four Connecticut Communities, 1980 - 1984*. Water Resource Investigations Report 89-4167. (Hartford, Conn.)

Walesh, Stuart, *Urban Surface Water Management*. John Wiley & Sons. (New York, 1989)

Whipple, W. et al., *Erosional Aspects of Managing Urban Streams.* Rutgers University Water Resources Research Institute. (New Brunswick, N.J., 1980)

CHAPTER SIX

Arnold, Chester L., and Gibbons, C. James, *Impervious Surface Coverage.* American Planning Association Journal, Vol. 62, No. 2. (Chicago, 1996)

Colston, Newton V., *Characterization of Urban Land Runoff.* American Society of Civil Engineers, National Meeting on Water Resources Engineering, Meeting Preprint 2135. (Los Angeles, 1974)

Connecticut Department of Environmental Protection, Bureau of Water Management, *1990 Water Quality Report to Congress.* (Hartford, Conn., 1990)

Connecticut Department of Environmental Protection, Water Compliance Unit, *Nonpoint Source Pollution: An Assessment and Management Plan.* (Hartford, Conn., 1989)

Field, Richard, et al., *Water Pollution and Associated Effects From Street Salting.* Journal of the Environmental Engineering Division of the American Society of Civil Engineers, Vol. 100, No. EE2. (New York, 1974)

Klein, Richard, *Urbanization and Stream Quality Impairment.* American Water Resources Association Bulletin, Vol. 15, No. 4. (1979)

Land-Tech Consultants Inc., *Carrying Capacity of Public Water Supply Watersheds: A Literature Review of Impacts on Water Quality From Residential Development.* Prepared for Litchfield Hills Council of Elected Officials. (Litchfield, Conn., 1989)

Murphy, James, *History of the Clean Water Act.* Connecticut Department of Environmental Protection Rivers Newsletter, Spring 1992. (Hartford, Conn.)

Roesner, Larry, editor, *Effect of Watershed Development and Management on Aquatic Ecosystems.* American Society of Civil Engineers. (New York, 1997)

Schueler, Thomas R., *Controlling Urban Runoff: A Practical Manual for Planning and Designing Best Management Practices.* Metropolitan Washington Council of Governments, Department of Environmental Programs. (Washington, D.C., 1987)

Schueler, Thomas R., *The Importance of Imperviousness.* Center for Watershed Protection. Watershed Protection Techniques, Vol. 1, No. 3. (Silver Spring, Md., 1994)

Stankowski, Stephen, *Population Density as an Indirect Indicator of Urban and Suburban Land Surface Modifications.* U.S. Geological Survey Professional Paper 800-B. (Washington, 1972)

U.S. Environmental Protection Agency, Water Planning Division, *Results of the Nationwide Urban Runoff Program, Vol. I – Final Report.* (Washington, D.C., 1983)

U.S. Geological Survey, *Effects of Urbanization on Peak Streamflows in Four Connecticut Communities, 1980-1984.* Water Resources Investigations Report 89-4167. (Hartford, Conn.)

CHAPTER SEVEN

Barton, James R., and Winger, Parley V., *Rehabilitation of a Channelized River in Utah.* In *Hydraulic Engineering and the Environment*, American Society of Civil Engineers. (New York, 1973)

Brookes, Andrew, *Channelized Rivers: Perspectives for Environmental Management.* John Wiley & Sons. (Chichester, England, 1988)

Gore, James A., editor, *The Restoration of Rivers and Streams: Theories and Experience.* Butterworth Publishers. (Boston, 1985)

Hammer, Donald A., *Creating Freshwater Wetlands.* Lewes Publishing. (Chelsea, Mich., 1992)

Hoggan, Daniel H., *Computer-Assisted Floodplain Hydrology and Hydraulics.* McGraw Hill Publishing Co. (New York, 1989)

Hogue, Robert W., *Stream Maintenance to Reduce Flooding.* Public Works, April 1981.

Hughes, William C., *Threshold Velocities in Erodible Soils.* American Society of Civil Engineers Preprint 3781. (New York, 1979)

Leopold, Luna B., *Hydrology for Urban Land Planning: A Guidebook on the Hydrologic Effects of Urban Land Use.* U.S. Geological Survey Circular 554. (Washington, D.C., 1968)

MacBroom, James Grant, *Applied Fluvial Geomorphology.* University of Connecticut, Institute of Water Resources bulletin, March 1981. (Storrs, Conn.)

Maryland Department of Natural Resources, Stormwater Management Division, *Standards and Specifications for Infiltration Practices.* (1984)

Miller, Jack G., and Tibbott, Ron, *Fish Habitat Improvement for Streams.* Pennsylvania Fish Commission. (Harrisburg, Pa.)

New York State Department of Environmental Conservation, Bureau of Water Quality, *Stream Corridor Management: A Basic Reference Manual.* (Albany, N.Y., 1982)

Peterson, Allan W., *Design of Mobile Boundary Channels.* National Symposium on Urban Hydrology and Sediment Control, University of Kentucky. (Lexington, Ky., 1975)

Rhode Island Department of Environmental Management, Office of Environmental Coordination, *Recommendations of the Stormwater Management and Erosion Control Committee Regarding the Development and Implementation of Technical Guidelines for Stormwater Management.* (1988)

Shield, Douglas, *Environmental Features of Flood Control Channels.* American Water Resources Association, Water Resource Bulletin, October 1982.

Smyth, Patrick, and Barnes, Jeffery, *Cooperation for Recreation and Streambank Retention.* In *Stormwater Management Alternatives*, University of Delaware Water Resources Center. (Newark, Del., 1980)

Stevens, Michael, Simons, Daryl, and Lewis, Gary L., *Safety Factors for Riprap Protection.* Journal of the Hydraulics Division of the American Society of Civil Engineers, May 1976. (New York)

Thomas, Carl H., Robbins, Samuel W., and McCandless, Donald E., *Fish and Wildlife Considerations in Small Reservoir and Channel Construction in the Northeast.* American Society of Civil Engineers, Annual Environmental Engineering Convention, Meeting Preprint 2388. (Kansas City, Mo., 1974)

U.S. Army Corps of Engineers, *Hydraulic Design Criteria*. (Vicksburg, Miss., 1971)

U.S. Fish and Wildlife Service, *Western Reservoir and Stream Habitat Improvements Handbook.* (Washington, D.C., 1978)

U.S. Department of Transportation, *Highways in the River Environment, Hydraulic and Environmental Design Considerations.* Federal Highway Administration Training and Design Manual. (Washington, D.C., 1975)

U.S. Environmental Protection Agency, *The Public Benefits of Cleaned Water: Emerging Groundwater Opportunities.* (Washington, D.C., 1977)

Vars, Charles R., *Wildlife Resources and Project Design Under New Federal Planning Initiatives.* In *Mitigation Symposium*, U.S. Department of Agriculture. (Fort Collins, Colo., 1979)

Walesh, Stuart, *Urban Surface Water Management*. John Wiley & Sons. (New York, 1989)

Whipple, W., et al., *Erosional Aspects of Managing Urban Streams.* Rutgers University Water Resources Research Institute. (New Brunswick, N.J., 1980)

Wingate, P.J., et al., *Guidelines for Mountain Stream Relocations in North Carolina*. North Carolina Wildlife Resources Commission Technical Report No. 1. (1979)

Wisconsin Department of Natural Resources, *Guidelines for Management of Trout Stream Habitat in Wisconsin.* Technical Bulletin No. 39. (Madison, Wis., 1967)

FOREMATTER REFERENCE

World Book, Inc., *The Word Book Encyclopedia*, Volume 16, p. 353. (Chicago, Ill. 1998)

GLOSSARY REFERENCES

American Society of Civil Engineers, *Glossary, Water and Wastewater Control Engineering.* (New York, 1981)

New York State Department of Environmental Conservation, Division of Water, *Stream Corridor Management.* (Albany, N.Y., 1986)

U.S. Geological Survey, *National Water Summary.* (Washington, D.C., 1987)

ADDITIONAL REFERENCES

Each of the topics in this book is covered by extensive literature not cited here. We encourage the reader to explore these topics in greater detail.

Index

M

N

O

T

U

V

W

Y

Designed by Mary Crombie, Acorn Studio,
Hartford, Connecticut
Using the Janson Text font family
Printed by
Hull Printing Company
Meriden, Connecticut
Mary Beth Uryga, Project Representative
Printed on Zanders Mega Dull recycled paper
Using vegetable-based inks
A Connecticut product.